Technical Writing Process 2nd Edition

TECHNICAL WRITING PROCESS

2ND EDITION

Master the Art of Technical Communication with Timeless Techniques and Modern Tools

ALISON PICKERING · AMANDA BUTLER CAITY CRONKHITE · FELICITY BRAND JOHN NEW · KIERAN MORGAN · STEVE MOSS · SWAPNIL OGALE

https://boffin.education/

Technical Details

ISBN: 978-0-9941693-2-7 (paperback)
Edition: 2nd Edition
Publication Date: February 2024
Authors: Alison J. Pickering, Amanda Butler, Caity Cronkhite, Felicity Brand, John New, Kieran Morgan, Steve Moss, Swapnil Ogale

Publisher's Information

Publisher: Boffin Education
Website: https://boffin.education

Copyright Statement: The content of this book, including all text, graphics, and other material ("Content"), is copyrighted by Boffin Education, unless otherwise indicated. All rights reserved. No part of this book may be reproduced, stored in a retrieval system, or transmitted in any form or by any means without the prior written permission of the publisher, except as provided by copyright law.

Templates: The templates available for download and linked to this book are for personal use and may be modified to suit individual requirements. These templates are also copyrighted material and may not be distributed, sold, or used for any commercial purposes without the explicit permission of Boffin Education.

Disclaimer: While every effort has been made to ensure the accuracy of the information contained in this book, Boffin Education and the authors expressly disclaim all liability for errors or omissions in this information and for any damages or losses resulting from the use of this material.

Trademarks: Any trademarks, service marks, product names, or named features mentioned in this book are assumed to be the property of their respective owners and are used only for reference. Boffin Education is not associated with any product or vendor mentioned in this book, and the publisher does not claim any association or ownership of such third-party materials.

Usage Limitations: This book is intended for educational purposes and should not be considered as professional advice or a substitute for consultation with qualified professionals. Use the content responsibly and in accordance with applicable laws and guidelines.

Updates or Corrections: Boffin Education reserves the right to make updates or corrections to this book. Any changes will be announced through suitable channels, which may include our official website or subsequent editions of the book.

Additional Notices: The views and opinions expressed in this book are those of the authors and do not necessarily reflect those of Boffin Education or any affiliated parties.

AI Training: The content of this book, including text, graphics, and all other material, is explicitly prohibited from being used for artificial intelligence training purposes or any form of data extraction and analysis intended to inform or contribute to AI systems, models, or algorithms. This restriction applies to all forms of AI, including but not limited to machine learning, natural language processing, and neural network training. Any violation of this restriction will be considered a breach of copyright and may result in legal action.

Praise for the First Edition

"**Every aspiring author / tech writer in college should have this book.**"

Steve Ballard, Manager, Enterprise Content Solutions

"**This is exactly the sort of guide I wish I had when I first started tech writing.**"

Sharon Witheriff, Technical Writer

"**The Technical Writing Process is a practical toolkit that demystifies the art of technical writing with useful templates to help you succeed quickly and smartly.**"

Richard Kidd, General Manager

"**Aspiring writers throughout the enterprise will find this to be a solid resource as they try to figure out how they should go about documenting something. It's logically laid out, not too heavy on jargon, and uses good diagrams and illustrations.**"

Duane Green, Senior Director of Content Strategy Solutions

"**I think this pragmatic approach is excellent and fills a desperate need in the tech writing world. Many a point raised will have the older hands nodding sagely and the younger people remembering the advice when they first run into those issues.**"

Dr. Charlotte Nash-Stewart, Engineer, Author, and Technical Writer

"**This is a well-written, comprehensive and practical guide for technical writers. It is packed full of great information and is very easy to read. I particularly like the 'Insights,' 'What does that mean?' and 'Tips' callouts which appear throughout the book.**"

Sue Geercke, Director Technical Communications

"Technical Writing Process explores the essential elements any technical writer should consider when tackling their next assignment. The book systematically guides the reader through an intuitive, yet effective, methodology that spans everything from Planning to Publishing.'

Shay Withnell, Product Marketing Manager

"Technical Writing Process is the professionalization of the technical writing discipline. It rightly considers the holistic context for a technical document and provides a comprehensive cookbook for getting it right within an organization. The insights, examples and templates are perfect and pragmatic. The definitive standard in technical writing."

Rami Banna, Product Lead

"This book serves as a great reference to someone tasked with the technical writing aspect of a project. It would have saved me much heartache and a considerable amount of time if I had this information at the start of my project. Your templates at the end of the book would have been really useful to me at the time."

Keo Phetsaya, Technical Sales Manager

Contents

About This Book 1
    Introduction 2
    Who This Book Is For 3
    The Process Model: Structured Learning 4
    Templates: Your Pathway into Practice 5
    Voice of Practitioner: Technical Writer Interviews 7
    Quality Assurance: Research and Peer Review 7
    About the Writers 8
    About Boffin Education 11
    How to Use This Book 12
    What's New 13
    Acknowledgements 15

Part 1 About the Profession 17

    Chapter 1 Introduction to Technical Writing 19
        1.1 Introduction 20
        1.2 What Is a Technical Writer? 21
        1.3 A Brief History of Technical Writing 23
        1.4 Joys and Frustrations of the Technical Writing Life 25

    Chapter 2 Roles and Responsibilities 27
        2.1 Introduction 28
        2.2 Types of Technical Writers 28
        2.3 The Docs They Write 30
        2.4 A Day in the Life of a Technical Writer 38
        2.5 Role Names 39
        2.6 Other Professional Writing Roles 40

    Chapter 3 Essential Skills 43
        3.1 Introduction 44
        3.2 Soft Skills 46
        3.3 Hard Skills 47
        3.4 Industry Knowledge 49

    Chapter 4 Breaking into Technical Writing 51
        4.1 Introduction 52

Contents

    4.2 Getting Qualified as a Technical Writer 54
    4.3 Continuing Professional Development 57
    4.4 The Application Process 57
    4.5 The Interview Process 73

Chapter 5 Career Growth and Survival **77**
    5.1 Introduction 78
    5.2 Transitioning to Your First Technical Writing Role 78
    5.3 Career Paths within the Industry 83
    5.4 Careers after Technical Writing 90

Chapter 6 Career Flexibility **93**
    6.1 Introduction 94
    6.2 Working across Multiple Time Zones 95
    6.3 Freelance Work 97

Part 2 Process *103*

Chapter 7 Tailor the Process **105**
    7.1 Introduction 107
    7.2 [Theory] The Technical Writing Process 107
    7.3 [Theory] Comparing Information-Development Frameworks 113
    7.4 [Practice] Customize the Process 115
    7.5 [Case Study] Case Study 1: Software Start-Up Company 118
    7.6 [Case Study] Case Study 2: Multinational Engineered Products Company .. 121

Part 3 Methods *127*

Chapter 8 AI for Technical Writers **129**
    8.1 Introduction 131
    8.2 [Theory] What Is a Large Language Model (LLM) and How Does It Work? .. 132
    8.3 [Opinion] Is Artificial Intelligence Going to Take My Job? 134
    8.4 [Example] Using ChatGPT to Document a Payment Software API 136

Part 4 Plan *147*

Chapter 9 Collect Information **149**
    9.1 Introduction 151
    9.2 [Theory] DIKW Pyramid 151
    9.3 [Practice] Collect Information, Data, and Knowledge 154

Chapter 10 Make a Plan **163**
    10.1 Introduction 165

10.2 [Theory] The PADRE Method	167
10.3 [Practice] Make a Plan	171

Chapter 11 Analyze Audience 179

11.1 Introduction	181
11.2 [Theory] Audience Types	182
11.3 [Theory] The Five Ws and One H	183
11.4 [Theory] Personas	187
11.5 [Practice] Analyze and Define Audience	189

Chapter 12 Define Review Team 193

12.1 Introduction	195
12.2 [Theory] Who's Who in a Review?	196
12.3 [Practice] Define Review Team	199
12.4 [Sample] Sample Review Matrix	203
12.5 [Sample] Sample Approval Matrix	204

Part 5 Design 205

Chapter 13 Design Structure 207

13.1 Introduction	209
13.2 [Theory] The Content Sandwich: Front Matter, Text, and Back Matter	210
13.3 [Theory] The Five Taxonomies	212
13.4 [Theory] Sequential and Nonsequential Taxonomies	219
13.5 [Practice] Design Structure	221
13.6 [Practice] Review and Iterate Structure	222
13.7 [Case Study] Cochlear™ Nucleus® 6 Sound Processor User Guide	223

Chapter 14 Design Stylesheet 227

14.1 Introduction	229
14.2 [Theory] Separation of Presentation, Content, and Structure	230
14.3 [Theory] Stylesheets	232
14.4 [Theory] Contrast, Alignment, Repetition, Proximity (CARP)	234
14.5 [Theory] Visual Patterns of Reading	242
14.6 [Practice] Design Stylesheet	244
14.7 [Sample] Sample Business Card Stylesheet	245

Chapter 15 Design Templates 253

15.1 Introduction	255
15.2 [Theory] Forms vs. Templates	256
15.3 [Theory] Metainformation, the Unique Characteristic of Templates	257
15.4 [Theory] Common Elements of Templates	258

Contents

    15.5 [Practice] Design Templates 259

    15.6 [Template] Technical Document Template 261

Part 6 Write 263

Chapter 16 Writing Principles 265

    16.1 Introduction 267

    16.2 [Theory] Empathizing with Your Audience 268

    16.3 [Theory] Organizing Information 271

    16.4 [Theory] Perfecting Prose 278

Chapter 17 Write Draft 287

    17.1 Introduction 289

    17.2 [Theory] The Stages of Drafting: First, Interim, and Final Drafts 290

    17.3 [Theory] The Write–Review–Update Cycle 291

    17.4 [Theory] Version Control 293

    17.5 [Practice] Write First Draft 296

Chapter 18 Include Images 305

    18.1 Introduction 307

    18.2 [Theory] Multimedia Learning: Your Brain Has a Dual-Core Processor 307

    18.3 [Theory] Cognitive Load Theory 310

    18.4 [Theory] Bringing It All Together: Cognitive Load and Technical Documentation .. 314

    18.5 [Practice] Crop, Capture, and Caption Images 315

    18.6 [Practice] Working with Graphic Designers 322

    18.7 [Practice] Translation Considerations 323

Part 7 Edit 325

Chapter 19 Edit Drafts 327

    19.1 Introduction 329

    19.2 [Theory] Levels of Editing 331

    19.3 [Theory] Editing Tools and Inputs 338

    19.4 [Practice] Edit Draft 342

Part 8 Review 347

Chapter 20 Review Draft 349

    20.1 Introduction 351

    20.2 [Theory] Write-Review-Update Process 352

    20.3 [Theory] Types of Review 353

    20.4 [Theory] Review Etiquette 355

20.5 [Practice] Review Draft 360

Chapter 21 Validate and Test Information 369
21.1 Introduction 371
21.2 [Theory] Factchecking 372
21.3 [Theory] Testing 372
21.4 [Theory] User Acceptance Testing (UAT) 374
21.5 [Theory] Usability Testing 375
21.6 [Theory] Informal Usability Testing 377

Part 9 Translate 379

Chapter 22 Translation Theory 381
22.1 Introduction 383
22.2 The Importance of Translation 383
22.3 [Theory] Translation, Localization, Internationalization 385
22.4 [Theory] Writing for Translation 387
22.5 [Theory] Setting Up Images for Translation 389
22.6 [Theory] Scope of Translation Services 390
22.7 [Theory] Roles and Responsibilities in the Translation Process 395

Chapter 23 Translation Practice 399
23.1 Introduction 401
23.2 [Process] Translation and Localization Process 401
23.3 [Practice] Translate and Localize Content 402
23.4 [Practice] Select Translation Partner 408
23.5 [Practice] Create Terminology Database 413

Part 10 Publish 415

Chapter 24 Publish 417
24.1 Introduction 419
24.2 [Theory] What Is a Document? 420
24.3 [Theory] Document Lifecycle 421
24.4 [Theory] Document Control 423
24.5 [Practice] Publish Final Version 430

Part 11 Manage 435

Chapter 25 Manage Progress 437
25.1 Introduction 439
25.2 [Theory] Checklists, Status Trackers, Visual Management Boards, and Schedules 441

Contents

25.3 [Practice] Manage Progress 452

Templates ***459***

Glossary ***465***

Index ***477***

References ***478***

About This Book

Lead Writer: Kieran Morgan | **Managing Editor**: Kieran Morgan

 Who Should Read This

- Aspiring Technical Writers
- Beginner Technical Writers
- Career Advancers
- Managers of Technical Writers
- Cross-Domain Professionals
- Educators of Technical Writers
- Consultants

CONTENTS

Introduction	2
Who This Book Is For	3
The Process Model: Structured Learning	4
Templates: Your Pathway into Practice	5
Voice of Practitioner: Technical Writer Interviews	7
Quality Assurance: Research and Peer Review	7
About the Writers	8
About Boffin Education	11
How to Use This Book	12
What's New	13
Acknowledgements	15

About This Book

Introduction

This is a book about technical writing—and how to do it like an expert. Technical writing is the art of creating instructions so that anyone can perform a task, whether that's as simple as using a hairdryer or as sophisticated as repairing a component in a satellite. Technical writing is a rapidly growing field that offers excellent salaries, great prospects for career progression and personal growth, and options for career flexibility.

We've written this book for technical writers—that is, the folks who write the instructions. It doesn't matter if you want to become one or you're already experienced in the job. It doesn't matter if you're weeks into your first technical writing job and riding the emotional rollercoaster of imposter syndrome, or if you have decades of experience. You'll still find something useful in this book.

The book is structured around the steps of the Technical Writing Process: Plan, Design, Write, Edit, Review, Translate, Publish, and Manage. We've also included some great information about the profession in Part 1: About the Profession. We interviewed technical writers at all career stages to gather insights from their experiences—from how they entered the field to what they think the future holds for the profession. To stay current, we include cutting-edge content in Chapter 8: Artificial Intelligence (AI) for Technical Writers, which highlights new techniques such as artificial intelligence.

In every chapter of this book, you'll find neat packages of information, from clear explanations of concepts that build your knowledge to practical steps and templates to apply your newfound knowledge in the real world. Each chapter is reviewed by experts and meticulously cross-referenced against leading industry and academic sources.

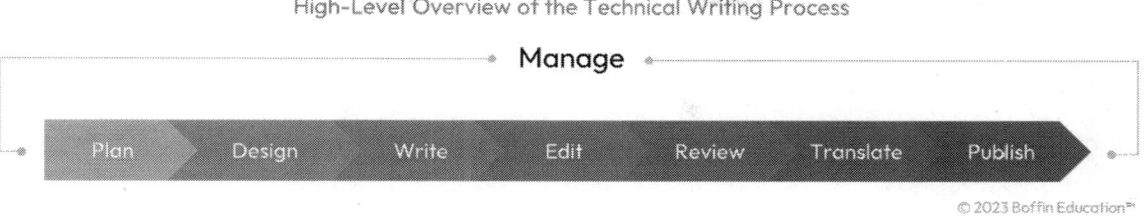

Figure 1: High-Level Overview of the Technical Writing Process

However, the true strength of this book lies in its creators. It was written, reviewed, and researched by technical writers, lecturers, and leading experts in their fields. We hope our passion for technical writing and the expertise of our collaborators shine through in these pages.

Good luck on your technical writing journey!

From the Writers

Who This Book Is For

Whether you're an experienced writer or still deciding on the right career, this book is for you. Specifically, this book caters to:

- **Aspiring Technical Writers** who are trying to land their first job and want an accessible guide that breaks down the basics of the profession—entry paths, qualifications required, what to expect on the job, different types and levels of seniority—and what tools to use to get your foot in the door, such as résumés, portfolios, and LinkedIn profiles.

- **Beginner Technical Writers** who don't have much experience and want to build their knowledge and skills to gain confidence in their work.

- **Career Advancers** who want to level up their skills and seek a promotion into a more senior role by deepening their craft or expanding their understanding of advanced skills.

- **Managers** who have been tasked with setting up teams and need an off-the-shelf framework for doing so.

About This Book

- **Cross-Domain Professionals** in other fields who have been tasked with doing something outside their area of expertise—technical writing—and need an easy-to-apply, ready-made framework, complete with templates.
- **Educators** who teach at the university or college level and are interested in adapting our process to develop courses for technical writers.
- **Consultants** who are seeking a systematic, process-oriented framework that offers downloadable and customizable templates and tools for client work.

If any of the above describes you, chances are you've chosen the right book. You're not alone—the Technical Writing Process has been successfully used by thousands of technical writers worldwide since 2015.

 What Does That Mean?

Technical Writer

A writer who develops (writes, edits, curates, etc.) technical documents. Also known as a technical communicator or technical author.

Technical Document

A document combining technical information with instructional guidance to help its audience accomplish a goal, such as carrying out a process or using a product. Examples include user guides, developer documentation, procedures, manuals, and quick-reference guides.

The Process Model: Structured Learning

The backbone of the Technical Writing Process is the process chapters. These chapters represent the macro steps that writers, such as yourself, undertake when crafting technical documentation. They're also the building blocks of professional competence: topics that enable you to learn and master your craft. They build your understanding of the concepts with the skills to apply them in real life.

Templates: Your Pathway into Practice

Each process chapter is based on the same framework: the Structured Content Process Model. They're divided into four main elements:

- **Theory**: concepts underpinning the craft
- **Practice**: practical steps to apply your theoretical knowledge
- **Templates**: ready-made tools for applying the theory in practice, ready to customize for your project
- **Samples**: examples and case studies of how theory has been applied in practice

These elements are worked into each chapter. The example below shows how the Structured Content Process Model can be used to explain any concept—from writing an instruction manual to cooking a meal.

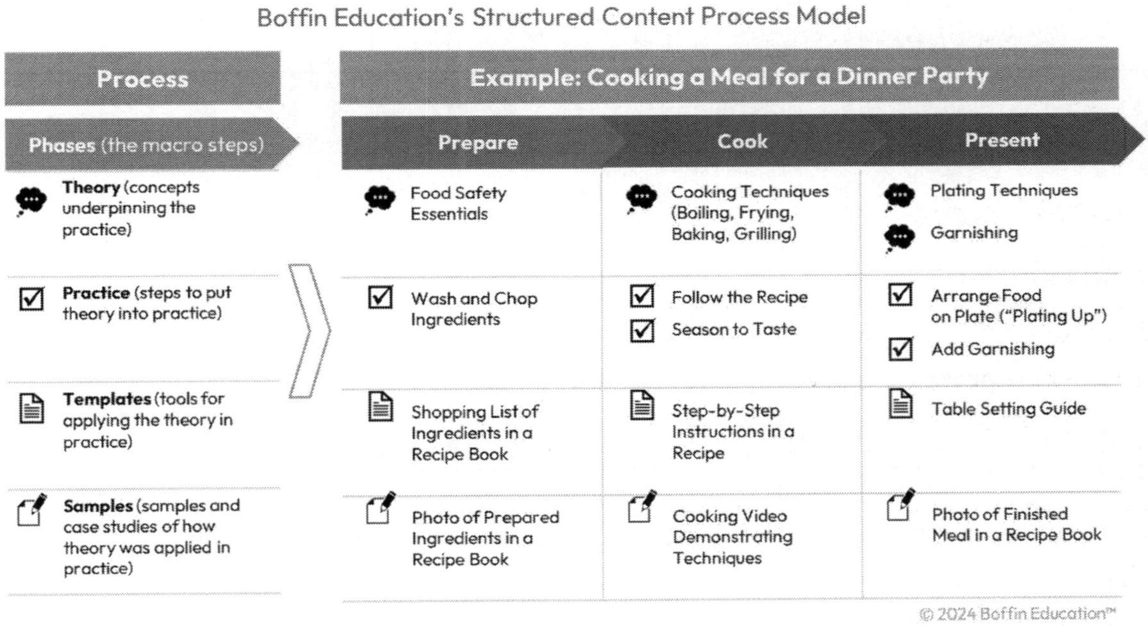

Boffin Education's Structured Content Process Model

Process	Example: Cooking a Meal for a Dinner Party		
Phases (the macro steps)	**Prepare**	**Cook**	**Present**
Theory (concepts underpinning the practice)	Food Safety Essentials	Cooking Techniques (Boiling, Frying, Baking, Grilling)	Plating Techniques; Garnishing
Practice (steps to put theory into practice)	Wash and Chop Ingredients	Follow the Recipe; Season to Taste	Arrange Food on Plate ("Plating Up"); Add Garnishing
Templates (tools for applying the theory in practice)	Shopping List of Ingredients in a Recipe Book	Step-by-Step Instructions in a Recipe	Table Setting Guide
Samples (samples and case studies of how theory was applied in practice)	Photo of Prepared Ingredients in a Recipe Book	Cooking Video Demonstrating Techniques	Photo of Finished Meal in a Recipe Book

© 2024 Boffin Education™

Figure 2: Boffin Education's Structured Content Process Model

Templates: Your Pathway into Practice

To facilitate the process, we've created templates designed to help you execute the detailed activities in each chapter of the Technical Writing Process.

About This Book

Templates Supporting the Technical Writing Process

- Agile Documentation Micro-Plan
- Audience Persona Template
- Briefing Checklist for Technical Writers
- ChatGPT Prompt Library
- Checklist for Using Visuals
- Documentation Plan Template
- Documentation Micro-Plan Template
- Editing Checklist Template
- Editing Sheet Template
- Estimating Sheet Template
- Resource Checklist
- Review Log Template
- Status Tracker Template
- Subject Matter Expert Interview Template
- Subject Matter Expert Workshop Template
- Technical Document Template
- Technical Writing Process Checklist
- Technical Writing Process Matrix
- Technical Writing Process Template
- Template Design Checklist
- Timeline Template
- Translation Best Practices Checklist
- User Guide Template
- Visual Design Principles Checklist

© 2024 Boffin Education™

Figure 3: Templates Supporting the Technical Writing Process

 **Note**

Some Templates Are for Subscribers Only

Some of the templates in this book are for subscribers only. You'll need to subscribe to our online knowledge base at https://boffin.education/ to access them in editable format. Purchasers of the e-book and paperback versions of this book can use the discount code in Templates to obtain a free one-year subscription and access the full breadth of our technical writing content. If you don't want to subscribe, head over to our website. There you'll find many of the more straightforward templates available for free, and others are presented in a noneditable format as images.

Voice of Practitioner: Technical Writer Interviews

To write this book, we interviewed more than twenty technical writers at all stages of their careers, from interns with only a month's experience to senior writers with over thirty-five years' experience who reflected on long and successful careers. We talked with documentation managers responsible for leading teams on different continents and aspiring technical writers wishing to break into the industry. We asked questions about every aspect of their careers, from the tools they use to their advice for newcomers to the profession and what they think the future holds. Snippets from these interviews are peppered throughout this book to provide valuable insights into the realities of the technical writing world.

Quality Assurance: Research and Peer Review

At Boffin Education, we're committed to the highest standards of research and review. Our content is meticulously researched and cited, as it draws on diverse practitioner and academic sources. We also follow stringent review standards. If you've read our content, you've probably noticed the following:

Lead Writer: Amanda Butler | **Peer Reviewer/s**: Felicity Brand, Kieran Morgan | **Expert Reviewer/s**: Saul Carliner

Every chapter must receive at least one review prior to publication. Most of our content is peer reviewed—that is, reviewed by a member of our writing team, all experienced practitioners. We also invite expert reviews from practitioners and academics who are leading experts in their field and, often, authors of journal articles and books that we've referenced.

This dedication to quality lends a trustworthiness to our content that you just won't find anywhere else.

About the Writers

The writing team for this edition of the *Technical Writing Process* comprises top-notch technical writers, specialists, and documentation managers from around the world. We've infused our passion for technical communication into this book, together with the latest advancements in the field and our best insights to help you succeed in your career.

Alison Pickering

Alison has more than twenty years of experience as a technical communications specialist and translation manager in Germany and the United Kingdom. She has a passion for creating high-quality content and a particular interest in developing and streamlining technical communications processes. Alison is based in the United Kingdom and enjoys riding horses and driving fast cars. Alison is the Lead Writer for Part 9: Translate.

Amanda Butler

Amanda has more than a decade of experience in technical writing. She is passionate about bringing new faces into the industry and specializes in managing globally distributed teams of technical writers. She can be found quoting Jane Austen and bad 1980s movies in Austin, Texas, where she lives with her husband and dog. Amanda is the Lead Writer for Part 1: About the Profession.

About the Writers

Caity Cronkhite

Caity is the founder and CEO of Good Words LLC, a technical writing and documentation consulting firm. Good Words delivers strategic, management, and implementation support for their clients' technical writing needs. Their writing wizardry and strategic skills streamline communications for a range of clients, from Fortune 500 companies to five-person start-ups. Check them out at http://www.goodwordswriting.com. Caity is Lead Writer for Chapter 8: Artificial Intelligence (AI) for Technical Writers.

Felicity Brand

Felicity has fifteen years of experience as a technical communicator. With a love for writing and a particular knack for editing, she enjoys working in open source where she can help writers (and nonwriters) create excellent written content. Felicity is based in Melbourne, Australia, and is a Lead Writer for Chapter 20: Review Draft.

John New

John has more than twenty-five years of experience as a technical writer. He especially enjoys creating templates and styles for organizations to help ensure high-quality structured documents. With a strong IT background, he enjoys tinkering with computers and solving twisty cubes. John lives in Sydney, Australia, and is a Lead Writer for Chapter 15: Design Templates, including the Technical Document Template and the Sample User Guide.

About This Book

Kieran Morgan

Kieran is the author of the *Technical Writing Process* (1st ed.). His experience spans documentation, leadership, operations, and project management. His company, Boffin Education, is dedicated to helping people like you become experts. Kieran lives in Sydney, Australia, with his partner and daughter. In addition to being Managing Editor for this book, Kieran is a Lead Writer for Part 2: Process, Part 4: Plan, Part 5: Design, Part 6: Write, Part 7: Edit, Part 8: Review, Part 10: Publish, and Part 11: Manage.

Steve Moss

Steve has more than three decades of technical communication experience. His notable contributions include spearheading educational content and leading an education department at WorkflowMax. An expert in online learning and a certified Information Mapping® instructor, Steve has also made his mark as a former president of http://TechComm.nz . He's based in Auckland, New Zealand, and is a Lead Writer for Part 6: Write.

Swapnil Ogale

Swapnil has more than sixteen years of experience as a technical writer. He consults with and works for a range of organizations to set up documentation teams, processes, workflows, and toolchains. This includes strategizing content needs, setting up information architecture, and facilitating user research for documentation sites. Swapnil enjoys speaking publicly about documentation and has presented at tech conferences and meetups across the globe. Based in Melbourne, Australia, Swapnil is a Lead Writer for Chapter 14: Design Stylesheet.

About Boffin Education

A "boffin" is British slang for an expert in their field. We like the sound of it, and it accurately captures the goal we share with our students: mastery of their chosen fields. We hope you like it too.

Boffin Education's mission is to help you become an expert by mastering theoretical concepts and learning the practical steps of your craft. Our products range from books and downloadable resources to online courses. We support newcomers to the field and established professionals looking for career growth.

We aim to prepare you for in-demand careers that offer you control over how you work:

- **Career Flexibility**: Find working styles that suit you: remote, hybrid, or in-office.
- **Work–Life Balance**: Fit modes of work to your lifestyle, including full-time, part-time, or freelance.
- **Attractive Remuneration**: Receive pay that is significantly higher than average.
- **Transferable Skills**: Gain skills that are highly transportable across multiple industries.

About This Book

All our content is written and reviewed by teams of "boffins" with deep experience in their fields. This ensures that it contains the best and most up-to-date information to assist you in becoming an expert.

How to Use This Book

You don't have to follow every step in the process. Its modular design and bite-size chapter format allow you to adapt the steps to your own projects. Chapter 7: Tailor the Process offers case studies and examples for modifying the process to suit your project. Other chapters cater to specialized needs. For example, Part 9: Translate is designed for those who need to localize their documents for different regions.

Symbols Used in This Book

While reading this book, you'll encounter reader aids, such as insights and tips, which are easily identifiable by their accompanying symbols. The following table explains the purpose of each.

Table 1: Symbols Used in This Book

	Insight	Insights offer an extra dimension to the guidance provided in this book. They may suggest career-enhancing moves, sound notes of caution, or shed additional light on a concept.
	What Does That Mean?	Definitions for technical jargon used in the accompanying text. All definitions are compiled in the glossary.
	Tip	Helpful advice for applying the concepts discussed in a particular section.

	Note Additional information about a concept or section that is not essential for understanding it but provides more context.
	Voice of Practitioner Anecdotes from individuals we interviewed during the creation of this book.
	Who Should Read This The intended audience that the authors had in mind when writing a chunk of content such as a chapter.

What's New

This edition isn't just an upgrade of *Technical Writing Process* (1st ed.)—it's a comprehensive overhaul. It refreshes much of the great content in the first edition and adds vastly expanded guidance from our new writing team. The structure has been revised to align with our Structured Content Process Model, which brings a better balance of theoretical knowledge and practical application to our content.

Table 2: What's New in the Second Edition of the Technical Writing Process

Part	What's Changed
About This Book	Streamlined introduction with a focus on the book's approach, process model, templates, and quality assurance.
Part 1: About the Profession	New section on the technical writing profession, including career insights and paths, based on global interviews with writers at various career stages.
Part 3: Methods	Updated information on generative AI in technical writing, with explanations of concepts and practical applications.

About This Book

Part	What's Changed
Part 2: Process	Realigned technical writing process with the Information-Development Life Cycle, including new phases and updated case studies.
Part 4: Plan	Revised planning content, enhanced Documentation Plan Template, including streamlined options, and increased focus on audience analysis.
Part 5: Design	Enhanced "Structure" section with guidance on taxonomies, stylesheets, and template design, including CARP concepts.
Part 6: Write	Expanded writing principles chapter, updated guidance on the write-review-update process, and new information on expert interviews and workshops.
Part 7: Edit	Made editing a stand-alone phase, with a modified version of the Levels of Editing concept.
Part 8: Review	Reworked testing and validation section, distinct from the review phase, and integrated with writing and editing through the process model.
Part 9: Translate	Added new guidance on translation and localization, including practical steps and theoretical concepts.
Part 10: Publish	Revised document control section, updated to align with international standards. Moved approval to this section.
Part 11: Manage	New content on managing technical documentation projects, including status trackers, checklists, kanban boards, and schedules.

Part	What's Changed
Templates	Updated templates, with integration of previously stand-alone tools into the documentation plan.
Samples	Added new section with practical samples illustrating applied concepts.
Glossary	Updated glossary with refreshed definitions and new terms.

Acknowledgements

The writers extend our heartfelt thanks to all those who have dedicated their time and effort to the successful completion of this project—particularly our expert reviewers, many of whom contributed guidance and support well beyond their chapter-by-chapter reviews. Your contributions and expertise have been invaluable. We'd also like to thank our partners and our families for their patience as we immersed ourselves in this project. We couldn't have done it without your support.

Expert Reviewers

- **Castella Arthur**: Chapter 18: Include Images
- **Dina Bennett**: Chapter 9: Collect Information, Chapter 13: Design Structure
- **Professor Saul Carliner**: Chapter 1: Introduction to Technical Writing, Chapter 2: Technical Writing Roles and Responsibilities, Chapter 3: Essential Skills for Technical Writers, Chapter 4: Breaking into Technical Writing, Chapter 5: Career Growth and Survival for Technical Writers, Chapter 6: Career Flexibility in Technical Writing
- **Patrick Lambe**: Chapter 13: Design Structure
- **Derek Moeller**: Chapter 8: Artificial Intelligence (AI) for Technical Writers
- **Stephanie Riches-Harries**: Chapter 22: Translation Theory, Chapter 23: Translation Practice
- **Emeritus Professor John Sweller**: Chapter 18: Include Images

About This Book

- **Deirdre Wilson**: Chapter 14: Design Stylesheet, Chapter 15: Design Templates
- **David Whitbread**: Chapter 14: Design Stylesheet

Collaborators

- **Sanja Kajfeš**, our fantastic graphic designer, who did the visual design for the book
- **John New**, who, in addition to being a Lead Writer, also contributed significantly to the book production through his mastery of templates and provided invaluable moral support during the journey
- **Kristen Hall-Geisler**, for her diligent copyediting and patience as we experimented with new ways of managing a writing project
- **Mischa Bendall**, who provided excellent administrative assistance during the writing
- **Ivana Devcic**, who contributed valuable feedback during the early stages of writing
- **Sue Geercke and the team at Cochlear** for their permission to use the *Cochlear™ Nucleus® 6 Sound Processor User Guide* as a case study in Chapter 13: Design Structure
- **Phil Cohen**, Director of HCi Professional Services https://hci.com.au/, provider of much-needed encouragement and advice, particularly on technical writer recruitment and testing

Interviewees

- Alison, Amanda, Anh, Annette, Ayo, Carly, Colin, Dina, Felicity, Francis, Jerome, John, Kate, Layale, Lee Anne, Nellie, Phil, Rachael, Robert, Sarah, and Swapnil

Part 1 About the Profession

Discover the fundamentals of technical writing, from its rich history to the diverse career opportunities it offers.

Chapter 1 Introduction to Technical Writing

Lead Writer: Amanda Butler | **Peer Reviewer/s**: Felicity Brand, Kieran Morgan | **Expert Reviewer/s**: Saul Carliner

Welcome to the rewarding world of technical writing, a profession that marries technical expertise with the art of clear communication. This chapter will welcome you into the universe of technical writers, providing insights into the role, the history of the craft, and the pathways into this career. Whether you're at the start of your career, contemplating a shift, or simply curious about the domain, the chapters in this part will provide you with a holistic view of what it means to be a technical writer. Dive in to discover if technical writing is for you.

 Who Should Read This

- Aspiring Technical Writers
- Beginner Technical Writers
- Cross-Domain Professionals

CONTENTS

1.1 Introduction 20
1.2 What Is a Technical Writer? 21
1.3 A Brief History of Technical Writing 23
1.4 Joys and Frustrations of the Technical Writing Life 25

1.1 Introduction

Technical writing can be an incredibly intellectually and emotionally satisfying, as well as financially rewarding, career. One of the great benefits of working as a tech writer is that as your reputation and experience grow, so do your options to work flexibly and craft your own work-life balance. Whether you're working on-site, remotely, or even venturing into freelance work, the world is your oyster.

This part discusses technical writing as a career. It explains what a technical writer does—and how to become one. We'll walk you through the process of applying for technical writing roles, discuss qualifications and skills, describe typical tasks, and detail possible career paths as well as flexible working options.

We'll share what to expect from a technical writing career—and help you figure out if that will suit you. Technical writing is different from almost all other forms of writing, whether that's fiction, marketing, or journalism. The point of technical writing is to teach your audience how to do a specific task, such as execute a process or use a product. Technical writing doesn't require specific qualifications or degrees, but it does require certain skills. If you want to improve your skills, we recommend the types of courses and classes to look for.

The application process for technical writing roles requires a résumé or LinkedIn profile, a cover letter, and writing samples. The hiring team uses these materials, a writing test, and interviews to evaluate your skills and determine whether you're a good fit for the role.

As you progress in your role, you'll have opportunities to further develop your skills. You'll get to choose between continuing to contribute as a technical writer, managing other writers, or branching into a related field, such as design, marketing, or product management. Each option provides a path for learning and growth.

Insight

Insider Perspectives: Technical Writer Interviews

When we wrote this book, we interviewed over twenty technical writers at all stages of their careers—from interns with only a month's experience to senior writers with over thirty-five years' experience who reflect on long and successful careers. We talked with documentation managers responsible for leading teams across continents and aspiring technical writers wishing to break into the industry. We asked questions about every aspect of their careers—from the tools they use to their advice for newcomers to the profession and what they think the future holds. You'll see snippets from these interviews used throughout the book, which we've collected in Interviews.

1.2 What Is a Technical Writer?

A technical writer is someone who explains technical concepts to an audience, traditionally through written text, and these days also through images, charts, and even videos. The goal is to help the audience understand the concept as quickly and fully as possible for the purpose of achieving some goal,[1,2] such as using a new gadget or software product. The goal is not to show off how smart you are by using complicated words—if you're trying to sound smart, you're doing it wrong! If you're trying to help your audience learn by using straightforward language that's appropriate for their level of education, you're doing it right.

The type of person who's drawn to tech writing is the type of person who's drawn to things that tech writers infuse in their work: a love of structure and information, a desire to empathize with their audience, and of course a technical focus. Great technical writers have an innate sense of curiosity that drives them (beyond monetary reward) and a passion for communicating their understanding to others. Technical writing, therefore, is a job for people who are passionate about building information and communicating that to others as knowledge.

If this sounds like you, you're in the right place—or at least reading the right book!

 Voice of Practitioner

John

- **Role**: Senior technical writer with more than 25 years of experience
- **Location**: Sydney, Australia
- **Work Mode**: 100% remote
- **Expertise**: Process, procedure, and software documentation

"I did a librarianship graduate diploma thinking I'd become a librarian. But I couldn't get a librarian job—in the early 1980s, jobs were a bit tight. So I got a job proofreading Braille books, using computers to produce Braille. Eventually I became the manager in the production unit, where I wrote documentation on how it worked. I thought, 'This is pretty interesting. How can I find a job that combines computers and writing?' And I found out there was a job called technical writing. It suited me—it had everything that I enjoyed doing, and for the last twenty-five years, technical writing has been the perfect job for me."

1.3 A Brief History of Technical Writing

Technical writing has roots as ancient as civilization itself. Wherever there were technical subjects needing explanation, some form of technical writing existed. The earliest humans created cave paintings and passed on oral instructions for making tools and hunting animals. Ancient civilizations such as the Aztecs, Babylonians, and Egyptians recorded numerous technical details—everything from the movements of the stars to lists of supplies needed for military campaigns. Great minds such as Aristotle, da Vinci, and Chaucer contributed to the field through history by turning their technical understanding into written knowledge and "infographics" passed on for the benefit of humanity.[3]

 Insight

Insight: How Ancient Is Technical Writing?

Geoffrey Chaucer, known for *The Canterbury Tales,* wrote a detailed technical guide to the astrolabe in the late 1300s, considered the first English technical manual.[4] This instrument was crucial for determining a ship's position at sea. About a century later, Leonardo da Vinci filled notebooks with sketches and detailed descriptions of innovative technical devices, one of which resembled a helicopter.[5] Not long after, pioneering European scientists such as Galileo, Robert Boyle, and Isaac Newton significantly influenced technical communication. They didn't just invent devices, such as Boyle's air pump, but also thoroughly explained their construction, usage, and the underlying theories.[6]

Technical writing today is a balance between science and storytelling, a combination of technical accuracy and persuasive prose. The best technical writers are capable of straddling both worlds—clearly communicating technical concepts through their mastery of language.

However, this wasn't always the case. In the early twentieth century, technical writing was mostly the domain of engineers and scientists. Their writings were seen not just as instructions, but contributions to the greater pool of humanity's knowledge, pushing progress ever onward. But in the first half of the twentieth century, with the turbulence that resulted from the technological progress and devastation of the two world wars, there was a rapid increase in new technologies that needed to be documented. Managers quickly realized the value of separating the "technical" from the "writing" and allowed engineers to focus their scarce time on science and engineering rather than writing reports[7].

Thus the modern profession of technical writing was born and the first official "technical writer" job titles appeared. Technical writing became more official in the 1950s with the formation of professional associations for technical writers, the establishment of technical communication journals, and later the creation of the first technical writing academic programs in the United States. The field exploded in the 1980s and 1990s when personal computers hit the market. The digital revolution changed the landscape of technical writing, which leaped from the printing press and bound-and-printed manuals to online help and video tutorials. The change exponentially increased the demand for technical writers[8].

For today's technical writers, that tension between the "technical" and the "writing" sometimes endures. The best technical writers combine a deep curiosity about the *technical* with a passion for the *writing* and embrace the richness of both worlds.

> **Voice of Practitioner**
>
> **Annette**
>
> - **Role**: Principal content strategist with 35+ years' experience
> - **Location**: California, United States
> - **Expertise**: Software documentation
>
> "When I started in '86, there wasn't a clear career path for technical writing. But now technical writing is critical—it's the translation layer between the software and the customer. I think tech writers tend to bury their light a little. I'd say carry yourselves with pride and be full partners in a cross-functional team. The software can't survive without the nexus that tech writers provide with the customer."

1.4 Joys and Frustrations of the Technical Writing Life

To give you a taste of the technical writing life, here are some joys and frustrations that many technical writers commonly experience.

Chapter 1 Introduction to Technical Writing

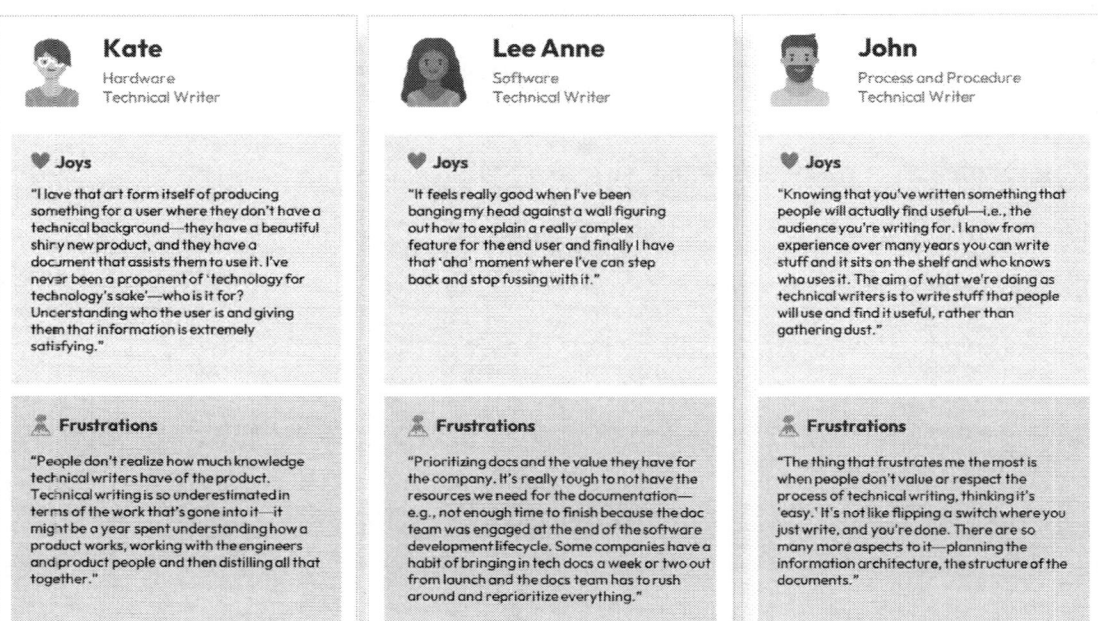

Figure 4: The Joys and Frustrations of Being a Technical Writer

Chapter 2 Roles and Responsibilities

Lead Writer: Amanda Butler | **Peer Reviewer/s**: Felicity Brand, Kieran Morgan | **Expert Reviewer/s**: Saul Carliner

This chapter demystifies the roles and responsibilities of technical writers by offering a look into their day-to-day activities and the range of specialties within the profession. You'll discover the different types of technical writers, understand the documentation they write, and gain insight into the value they bring to organizations. Furthermore, this chapter will touch upon job titles and professional roles that share a foundation with technical writing.

 Who Should Read This

- Aspiring Technical Writers
- Beginner Technical Writers
- Cross-Domain Professionals

CONTENTS

2.1 Introduction	28
2.2 Types of Technical Writers	28
2.3 The Docs They Write	30
2.4 A Day in the Life of a Technical Writer	38
2.5 Role Names	39
2.6 Other Professional Writing Roles	40

2.1 Introduction

If you're an aspiring or new technical writer, understanding the options available to you as you contemplate your future career is helpful. Like many other professions, technical writing offers several specializations, some of which might be more suited to your skills, experience, and personal preferences than others—and some of which you might not have known existed. This chapter will help you understand the different specializations and give you a taste of what you can expect in each. We'll take a deep dive into each specialization, exploring the type of documents they create and who their main audience is.

Whether you're crafting user-friendly manuals for hardware, getting to grips with a new software product, or interviewing a subject matter expert to create accurate process and procedure documentation, the role of a technical writer is as varied as it is essential. Technical writers are a critical bridge between technology and the end user, helping to distill complex concepts into usable prose.

Finally, this chapter will explore the horizon beyond traditional roles to examine the evolving scope for employment in the field and related professional writing opportunities. As the landscape of technical communication continues to shift and expand, so too do the opportunities for those skilled in the art of conveying complex information with clarity.

2.2 Types of Technical Writers

There are several categories of technical writer. The tasks and responsibilities will depend on the type of work you are doing:

- **Hardware technical writers** create online and printed (and sometimes translated) manuals, typically for end-customer use in manufactured goods.

- **Software technical writers** document user interfaces for end users. They typically have some coding experience and understand the software development lifecycle and its tools. They often use structured markup such as Darwin Information Typing Architecture (DITA).

- **Process and procedure technical writers** create documentation for an audience that is mostly within (i.e., internal to) the organization they work for. This includes process and procedure-driven documents such as operations manuals, business processes, user guides, and work instructions.

 Insight

Developer Documentation Technical Writers

A relatively new specialization for software technical writers is the developer documentation technical writer. They are a more "technical" breed of technical writer, as amateur coding skills are a minimum requirement. They follow a "docs-as-code" approach by integrating their documentation into the software code written by developers to create records such as API documentation for a developer audience. They have a strong understanding of the software development lifecycle.

Technical writers will often specialize in one of the above categories, as the toolsets and skills tend to be distinct and often carry a steep learning curve. The necessary soft skills [see below] are of course common across all technical writing types.

Chapter 2 Roles and Responsibilities

The Main Technical Writer Job Types—Hardware, Software, and Process and Procedure

Kate
Hardware
Technical Writer

"I write instructions for medical devices. I'm passionate about understanding the audience's needs and communicating that clearly. I've never considered myself a 'techy' technical writer."

Lee Anne
Software
Technical Writer

"I document software for an engineering software company. I love writing, tech, and working with people. My passion is information architecture and tinkering around with coding languages in my spare time."

John
Process and Procedure
Technical Writer

"I document operational processes for a telecommunications company. I enjoy developing relationships with subject matter experts and writing clearly. Being an expert in certain software is good, but it's not the be all and end all."

© 2023 Baffin Education™

Figure 5: The Main Technical Writer Job Types—Hardware, Software, and Process and Procedure

2.3 The Docs They Write

The different types of technical writers create distinct documentation types for different audiences. We've provided some basic definitions and examples below.

2.3 The Docs They Write

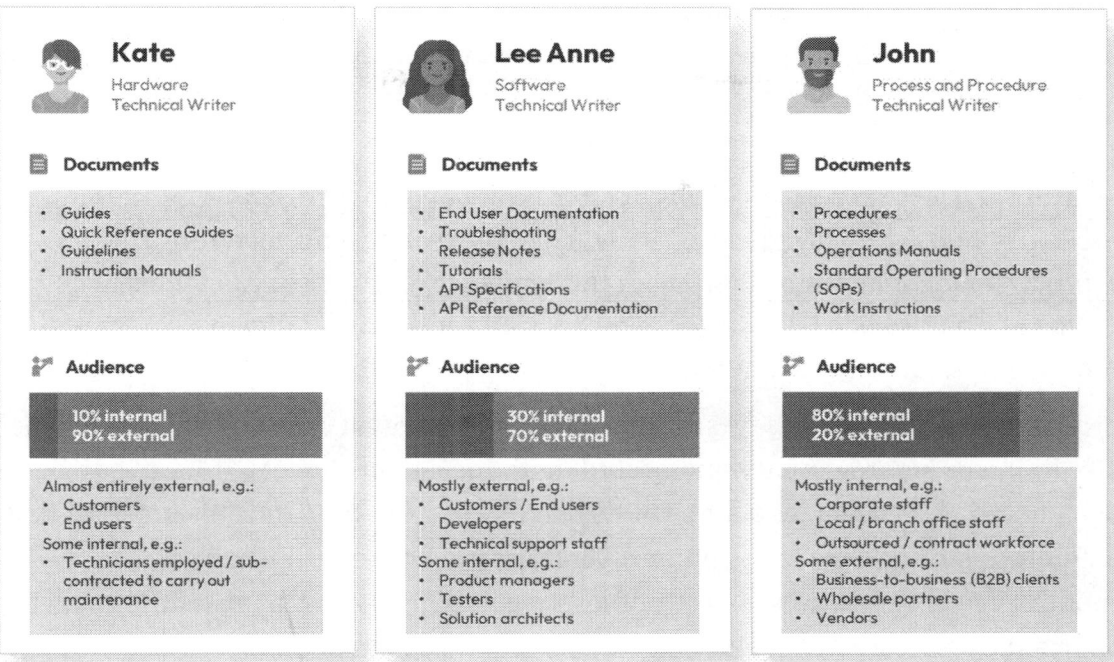

Figure 6: Types of Documents Written by Technical Writers

2.3.1 Hardware Documentation

Hardware documentation is a type of technical documentation that explains the use or design of a hardware device and its components. This typically includes explanations of its user interface, normal operating limits and specifications, care and maintenance, troubleshooting if something goes wrong, and any safety hazards the customer should be aware of while using the device.

The type of hardware being documented might include anything from personal computers and phones, which have increasingly minimalistic documentation, to complex industrial devices that have nonintuitive interfaces, extensive customization, and troubleshooting that trained operators have to master.

Hardware documentation can be for:

- **External users** such as customers, end users, or partners using the device. External documentation—particularly that shipped with the device—can be high-profile documents. It often combines the need to help the customer quickly and easily understand a product with the necessity of being a polished piece of visual design that strongly reflects the brand of the company.

- **Internal users** such as engineers, testers, product managers, and subject matter experts who are involved in designing, building, testing, and validating the product. Documentation written for this audience (such as specifications) will typically have much less of a focus on branding and presentation and more of a focus on accurately communicating technical information and specifications.

Hardware documentation is often subject to stringent regulatory controls, which may differ from country to country and according to the risk profile of the hardware being documented. For example, a surgical device used in the operating theater will have much more stringent controls on its documentation than a mechanical dog-feeder. Technical writers need to be aware of these regulations. They may mandate small but important details, such as a particular shape of symbol that must be used for warning signs or electrical hazards and particular marks, such as the "CE" symbol that indicates that a product has been approved for use within a certain region. When in doubt, you check with a more senior tech writer on your team or your organization's regulatory folks.

Examples: Guides, instruction manuals, quick-reference guides, guidelines, specifications.

2.3.2 Software Documentation

Software documentation is a type of documentation that explains the use and design of software and its components. This typically includes explanations of its user interface, how to install it, how to configure it for a customer's needs, and troubleshooting it if something goes wrong. Even software that you might think is fairly straightforward, such as email, will have extensive documentation and troubleshooting.

Software documentation can be for:

- **External users** such as customers, end users, or partners using the software. External documentation is usually focused on how to use, install, configure, and troubleshoot the software. These days, it's most often published on a company's website or external knowledge base, or it's integrated into the software itself. The focus is on clarity of instructions, responding to emerging customer questions as they emerge, and keeping the documentation current as new features are released.

- **Internal users** such as software developers, architects, testers, product managers, and subject matter experts who are involved in designing, building, testing, and validating the software. Documentation for this audience will typically have a focus on the software's internal architecture.

Examples: User guides, user manuals, end-user documentation, troubleshooting, release notes, specifications.

2.3.3 Developer Documentation

Developer documentation is a specialized type of software documentation. It helps developers understand how to develop, integrate, and customize software products such as application programming interfaces (APIs) and software development kits (SDKs).

This is a relatively new type of documentation that requires technical writers to have some understanding of programming languages and their associated web tools, such as GitHub, and software engineering processes, such as the software development lifecycle (SDLC).

Developer documentation can be for:

- **External users** such as customers, partners, and developers who are using or integrating with the software.
- **Internal users** such as developers, testers, architects, and technical product managers who are involved in creating the software.

Chapter 2 Roles and Responsibilities

Technical writers who work in this area can also be known as programmer writers or API technical writers.

Examples: API documentation, SDK documentation, reference guides, integration guides.

2.3.4 Process and Procedure Documentation

Process and procedure documentation includes any document that assists someone in carrying out a process or procedure, whether or not that involves a software component. It might even be a completely manual (nonautomated) business process. Processes can range from something as simple as logging a ticket with your building's maintenance team for a broken lightbulb to something as complicated as swapping out a nuclear reactor on a submarine.

Process and procedure documentation can be for:

- **External users** such as vendors, partners, or business-to-business clients of the organization.
- **Internal users** such as employees; contractors and others within the organization's workforce; retail, local, or branch office staff; a field workforce employed to carry out installations, inspections, servicing, and so on, for the organization beyond the corporate headquarters; and managers who need to ensure policies are implemented consistently and updated continually to reflect best practice.

Process and procedure documentation typically focuses on clearly communicating the organization's agreed-on best-practice method and sequence of carrying out a task, or using an internal software system. It's typically published via an organization's branded templates, whether that's in Microsoft Word or a help authoring tool (HAT)'s templated output.

Examples: Procedures, processes, flowcharts, operations manuals, standard operating procedures (SOPs), work instructions.

2.3.5 Internal vs. External Documentation

In technical writing, documentation is typically categorized as either for internal or external users, each serving distinct purposes and audiences.[9]

- **External documentation** is created for users outside the organization. These users include customers, end users, or partners who interact with the organization's products or services. This documentation often takes the form of user guides, instruction manuals, and troubleshooting resources. It is designed to be accessible and user friendly, focusing on helping the user understand and effectively use a product or software. External documentation is also a reflection of the organization's brand and values, often requiring a polished visual design and clear, concise language.

- **Internal documentation**, on the other hand, is intended for an audience within the organization. This audience could be employees, contractors, or other stakeholders involved in the design, development, testing, and maintenance of products or services. Examples of internal documentation include technical specifications, process guidelines, and development guides. This type of documentation is more technical and detailed, focusing on accurately communicating internal processes, technical details, and procedural information. It is less concerned with branding and visual design and more focused on precision and clarity of technical content.

Both types of documentation are crucial in their contexts. External documentation serves as a bridge between the product and its users, enhancing user experience and satisfaction. Internal documentation plays a vital role in maintaining operational efficiency and ensuring that products are developed and maintained according to the required standards and specifications.

Understanding the distinction between these two types of documentation is essential for technical writers, as it influences the style, content, and presentation of the information they create.

> **What Does That Mean?**
>
> **Internal Documentation**
>
> Technical content for an organization's internal audience, focusing on technical details, processes, and operational efficiency.
>
> **External Documentation**
>
> User-oriented content for customers and partners outside the organization, emphasizing ease of use, product understanding, and brand representation.

2.3.6 What's Driving the Need for All This Documentation, Anyway?

Fundamentally, technical documentation is something that helps a user understand how to carry out a task, whether that's using a software system or a gadget they've just bought. If the software or hardware isn't 100 percent intuitive—that is, the user can't figure it out just by using it—then documentation will often be their first port of call.

For any organization that doesn't want to be flooded by calls to its technical support team, making good, clear documentation easily available for their products is crucial. It helps divert technical support calls, and it mitigates any usability problems with the product that can't easily be designed away. This is even more important if the product is specialized or complex.

The same goes for process and procedure documentation. As organizations grow larger, they develop their own unique processes and procedures to carry out work in the way they believe is best for their customers. They also want to help their organization save money by doing things the most efficient way. By creating documentation on the best way to carry out these tasks, they accomplish several important goals:

- **Standardization**: ensuring tasks are consistently executed
- **Scalability**: preserving the intended ways of working, even as more employees join
- **Efficiency**: streamlining operations while building on past lessons

Sometimes the need for documentation is driven by external regulations or best-practice frameworks, such as ISO 9001. These frameworks and regulations dictate to the organization which type of documentation it needs—for internal or external audiences, or both. These frameworks may impose extensive hierarchies of documentation that the company needs to create and maintain to comply, and it creates jobs for technical writers like you (if that's your thing).

 Voice of Practitioner

Jerome

- **Role**: Technical Writer with 4+ years' experience and 10+ years' language and writing tutoring
- **Location**: Washington, United States
- **Expertise**: Software documentation

"I think the biggest misconception of tech writers is that they're expendable in an organization and don't provide much value. Docs are among people's first experience with a product. That initial impression makes a reputation faster than marketing."

2.4 A Day in the Life of a Technical Writer

It may surprise you that writing is only one aspect of technical writing. You can only start writing when you understand whom you're writing for and what you're writing about. In addition to research, interviews, and writing, technical writing involves collaboration through planning, editing, and more. As your projects progress, as a technical writer, you'll move from one phase of the technical writing process to another—or more likely, move back and forth between phases as you juggle numerous documentation tasks.

Below is an example of a typical "day in the life" of the different technical writer types, showing all the different tasks they do in their day to achieve their end goal: writing great documentation!

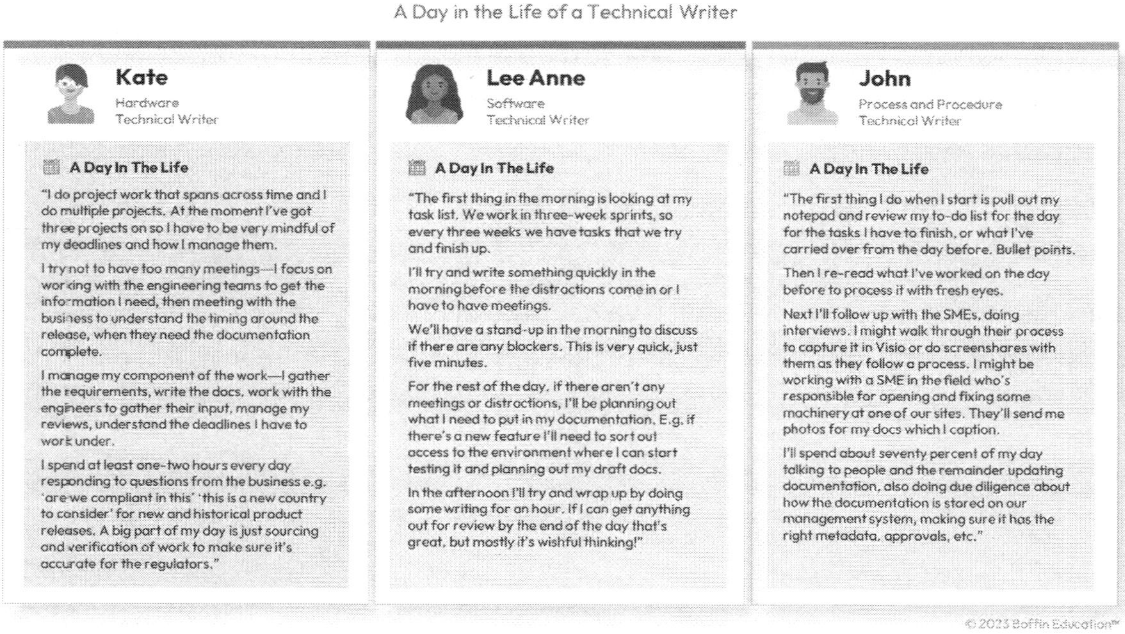

A Day in the Life of a Technical Writer

Kate — Hardware Technical Writer

A Day In The Life

"I do project work that spans across time and I do multiple projects. At the moment I've got three projects on so I have to be very mindful of my deadlines and how I manage them.

I try not to have too many meetings—I focus on working with the engineering teams to get the information I need, then meeting with the business to understand the timing around the release, when they need the documentation complete.

I manage my component of the work—I gather the requirements, write the docs, work with the engineers to gather their input, manage my reviews, understand the deadlines I have to work under.

I spend at least one-two hours every day responding to questions from the business e.g. 'are we compliant in this' 'this is a new country to consider' for new and historical product releases. A big part of my day is just sourcing and verification of work to make sure it's accurate for the regulators."

Lee Anne — Software Technical Writer

A Day In The Life

"The first thing in the morning is looking at my task list. We work in three-week sprints, so every three weeks we have tasks that we try and finish up.

I'll try and write something quickly in the morning before the distractions come in or I have to have meetings.

We'll have a stand-up in the morning to discuss if there are any blockers. This is very quick, just five minutes.

For the rest of the day, if there aren't any meetings or distractions, I'll be planning out what I need to put in my documentation. E.g. if there's a new feature I'll need to sort out access to the environment where I can start testing it and planning out my draft docs.

In the afternoon I'll try and wrap up by doing some writing for an hour. If I can get anything out for review by the end of the day that's great, but mostly it's wishful thinking!"

John — Process and Procedure Technical Writer

A Day In The Life

"The first thing I do when I start is pull out my notepad and review my to-do list for the day for the tasks I have to finish, or what I've carried over from the day before. Bullet points.

Then I re-read what I've worked on the day before to process it with fresh eyes.

Next I'll follow up with the SMEs, doing interviews. I might walk through their process to capture it in Visio or do screenshares with them as they follow a process. I might be working with a SME in the field who's responsible for opening and fixing some machinery at one of our sites. They'll send me photos for my docs which I caption.

I'll spend about seventy percent of my day talking to people and the remainder updating documentation, also doing due diligence about how the documentation is stored on our management system, making sure it has the right metadata, approvals, etc."

© 2023 Boffin Education

Figure 7: A Day in the Life of a Technical Writer

2.5 Role Names

Technical writers aren't always called technical writers. In fact, there are many names for our profession, and they vary between countries and even organizations. As experienced technical writers will tell you, some prefer to be called technical communicators, or information architects—the list goes on! Professional bodies in the field are usually called societies for technical communication. The broader term embodies what we do; after all, a key point of this book is that technical writing isn't just about the writing. It doesn't always involve the creation of written documents; technical writers may create other forms of content such as e-learning modules, presentations, video tutorials, and technical illustrations. They may even create material with minimal or no text—for example, guides that rely almost exclusively on images rather than words.

Some other names for technical writers include:

- Content designer
- Content strategist
- Digital storyteller
- Documentation manager
- Documentation specialist
- Information architect
- Technical author
- Technical communicator
- Technical editor
- Technical evangelist

> **Insight**
>
> Because technical writing roles can have so many different names, if you only search for "technical writer," you could limit your potential opportunities. For a broader search, try mixing it up with some of the other job titles shown above. You may find that in your country, another job title is equally as common, or even more so.

2.6 Other Professional Writing Roles

There are numerous opportunities for professional writers beyond technical writing. Although these roles might not be called technical writers, many of the core hard and soft skills overlap with technical writing. All of them involve a core element of writing and editing competence. If you've built up your skills in one professional writing career, great news: you may find it very easy to transition to another with a little additional training or experience.

Here are some examples:

2.6 Other Professional Writing Roles

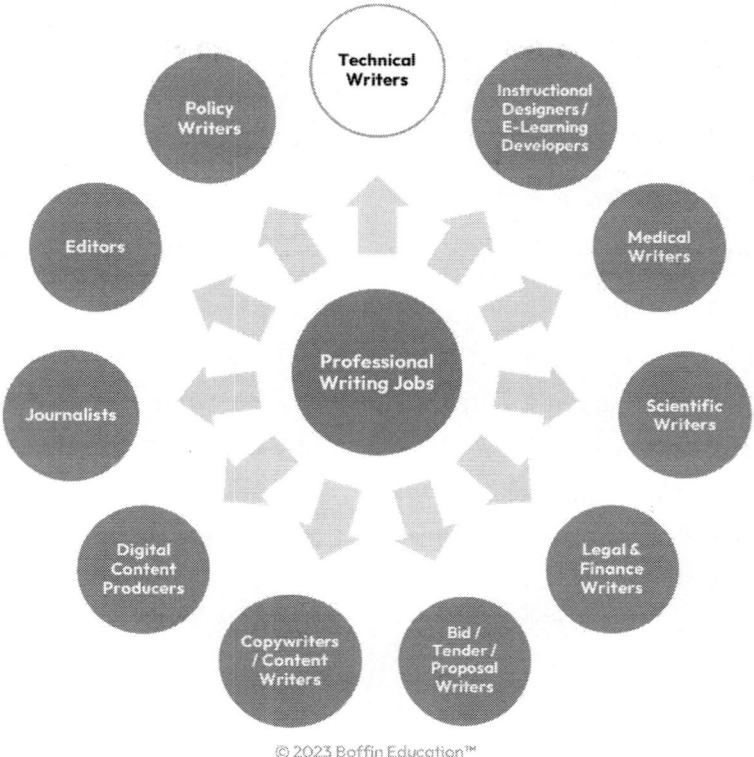

Figure 8: Professional Writing Jobs Beyond Technical Writing

Chapter 2 Roles and Responsibilities

 Voice of Practitioner

Francis

- **Role**: Technical writer with 4 years' experience
- **Location**: Manchester, United Kingdom
- **Expertise**: Software documentation

"I was working in an engineering manufacturing company, where I started as a trainee production operator. I was fixing machines for the automotive industry, doing the physical tooling—it was a hands-on job. During that time I went back to uni on a scholarship with the company to do a STEM degree. In studying, a job came up as a technical writer at my company, and when I applied, they thought I was a good fit because I knew the company, I knew the tools, and I was proving to be literate—I was quite a company guy."

Chapter 3 Essential Skills

Lead Writer: Amanda Butler | **Peer Reviewer/s**: Felicity Brand, Kieran Morgan | **Expert Reviewer/s**: Saul Carliner

This chapter provides a comprehensive guide to what it takes to excel in a technical writing career, emphasizing the interplay between hard and soft skills. It discusses the importance of understanding new technologies and effectively communicating them through writing while also highlighting essential interpersonal skills for working with subject matter experts.

 Who Should Read This

- Aspiring Technical Writers
- Beginner Technical Writers
- Cross-Domain Professionals

CONTENTS

3.1 Introduction	44
3.2 Soft Skills	46
3.3 Hard Skills	47
3.4 Industry Knowledge	49

3.1 Introduction

If you're considering jumping into a technical writing career, you need to consider whether the day-to-day life of a technical writer is for you and whether you have the kind of skills that technical writers need to excel in their role, or the aptitude to learn them. Not everyone is suited to be a technical writer.

Technical writing, at its heart, requires you to be able to learn new technologies and communicate them effectively to others through your writing. You must be able to understand technical information well enough to explain it to others, and you must be able to write clearly in a way that other people understand. To obtain the technical information in the first place, you need to be able to work with subject matter experts who hold this information in their minds. Interpersonal skills are an indispensable element of any successful technical writer's toolkit.

Some folks truly love technical writing—it's in their bones. Others may consider the role a stepping stone to another career.

3.1 Introduction

Essential Soft and Hard Skills for Technical Writers

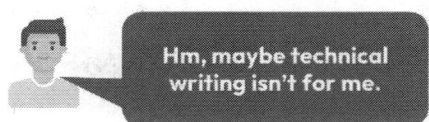

✓ **Empathy:** A technical writer should always seek to understand their audience's perspective and anticipate possible questions they may have.

✓ **Thirst for Learning:** A technical writer should always seek to gain a deep understanding of the subject matter to convey their understanding to the audience—and to gain respect from their subject matter experts.

✓ **Interpersonal Skills:** Building strong relationships with subject matter experts is essential for technical writers to get the information they need.

✓ **Patience:** A technical writer needs to be patient and adapt to their subject matter experts' pace and preferred style of work to build trust and respect.

✓ **Communication Skills:** Strong written and verbal communication is essential for a technical writer to get their point across in their documentation and to communicate with subject matter experts.

✓ **Time Management Skills:** A technical writer needs to be able to manage review cycles and push back on unrealistic deadlines politely and professionally.

✗ **Lack of Regard for the Audience:** Failing to consider the targeted audience may lead to confusing and unclear documentation.

✗ **Lack of Curiosity for the Subject Matter:** A lack of interest in understanding complex subject matter can result in poor documentation or require subject matter experts to spend their scarce time making numerous corrections.

✗ **Lack of Attention to Detail:** Technical writers need a passion for getting the finer points right. A seemingly trivial detail might be critical for the user to successfully carry out a task.

✗ **Poor Writing Skills:** Poor writing skills will result in substandard documentation. Great technical writers understand the nuances of the right word choice and its implications for the audience, as well as possess more technical skills like reviewing, editing, and proofreading.

✗ **Closed to Receiving Feedback:** Being closed to receiving constructive feedback will hamper the quality of your work. Receiving and processing feedback from colleagues, subject matter experts, and others is a key stepping stone on the path to writing excellence.

© 2023 Boffin Education™

Figure 9: Essential Soft and Hard Skills for Technical Writers

 Insight

Soft Skills vs. Hard Skills

Soft skills—also known as people skills or interpersonal skills—are the behaviors and attitudes you'll need to bring to your work so you can flourish as a technical writer. Hard skills are technical skills that are easily measurable, i.e., you can prove them with a degree or certificate. Both can be learned, and both are essential to your credibility as a technical writer, but it's much easier to prove on paper that you have hard skills. As your career as a technical writer grows, aim to gain a reputation among employers for your soft skills as well as your technical excellence.

3.2 Soft Skills

The soft skills you need to be a great technical writer tend to be consistent across organizations. Just like hard skills, you can learn soft skills by taking a course; asking for feedback from a manager, coach, or mentor; or reflecting on how you handled a situation.

Important soft skills for technical writers include:

- Working well with subject matter experts.
- Attention to detail.
- Ability to quickly learn new technical concepts.
- Ability to empathize with your audience.
- Ability to manage your own time.

Great technical writers are known as much for their soft skills as for their technical proficiency, and over time they develop a reputation in the industry for their winning combination of soft and hard skills. These technical writers are almost always in high demand and can usually command top-of-the-market rates, take their pick of roles with great organizations, or negotiate employment conditions that work for them.

Insight

Use Reflections to Build Your Soft Skills

Reflecting on how you handled a situation can be a great pathway to personal growth. Take the opportunity to reflect on a situation you were involved in, either at work or in your personal life, that you think went particularly well—or not so well, providing you with an opportunity to do better next time. Keep your reflections filed somewhere you can add to them over time. Reflections can be a great way of identifying behaviors that have worked well—for yourself or for others—and incorporating them into your repertoire of soft skills. You can also use them to identify habits that you think might need changing. Don't beat yourself up if you haven't handled a situation as well as you could. Reflections help you understand what you can do better next time.

3.3 Hard Skills

The hard skills you need depend on the type of technical writing you do. These skills range from the geeky end of the technical writing spectrum, where employers in high-tech software industries will have an expectation that technical writers can code—at least to some extent—to roles that require very little specialized technical knowledge beyond the standard technical writer toolkit, such as documenting processes or procedures.

Some hard skills every technical writer must have, regardless of the organization they work for, are the abilities to:

- Effectively structure and format documents.
- Write clearly and concisely.
- Use correct grammar and punctuation.
- Follow a style guide.

Chapter 3 Essential Skills

Here are some of the skills and tools commonly needed for different technical writing roles:

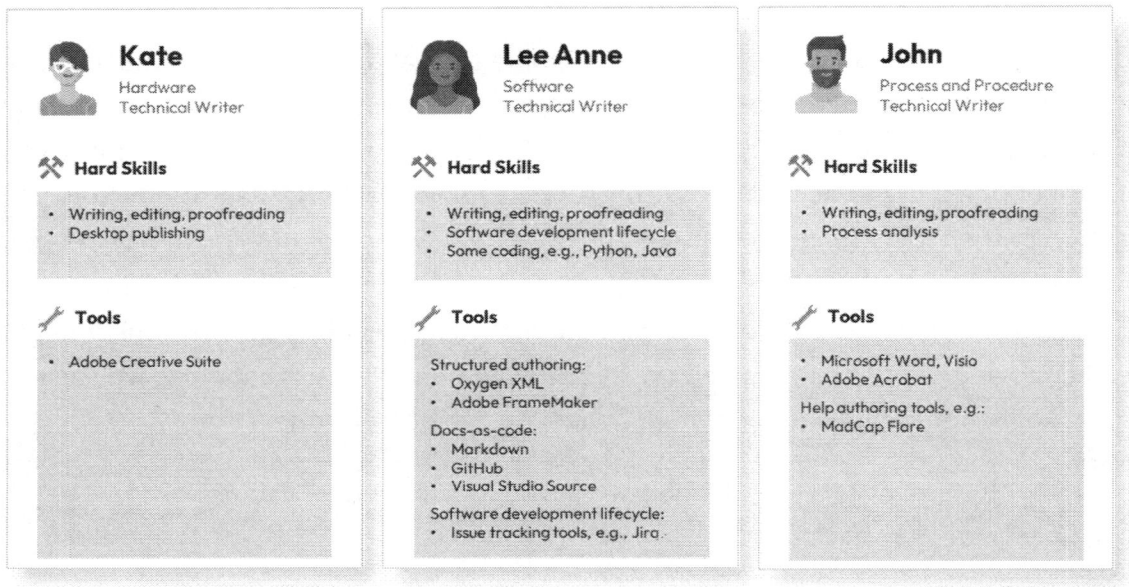

Figure 10: Hard Skills and Tools Required for Different Technical Writing Roles

Note

There are many more hard skills that are highly desirable for technical writers to develop. They include:

- Degrees and certificates
- Understanding of fundamental technical writing concepts
- Proficiency with writing, grammar, and punctuation
- Knowledge of common rules of style and grammar
- Ability to effectively use a style guide
- Ability to effectively structure and format documents
- Proficiency in a computer programming language

3.4 Industry Knowledge

If you spend time working in a particular industry, you'll notice that folks around you use what's called industry jargon—that is, unique terms, acronyms, and concepts unique to that industry and impenetrable to anyone new to it. If you're new to the industry, you'll need to learn this industry-specific knowledge on the job. Your first six months might be slightly bewildering as the people around you seem to speak in code, but bear with it. It's worth spending the time coming to grips with the terminology and technology of your industry, as well as any legal requirements particular to it that govern how you write your technical documents.

Here are some examples of industries where documentation is governed by regulations:

- **Aviation industry**: US Federal Aviation Administration (FAA) requirements for documentation

- **Medical device industry**: US Food and Drug Administration (FDA) requirements for documentation

As you can imagine, more technical or tightly regulated industries come with more complex jargon and regulations. However, technical writers who invest time developing a deep, industry-specific knowledge often develop a reputation as an expert writer in that industry. If that's you, expect to be in high demand!

Chapter 4 Breaking into Technical Writing

Lead Writer: Amanda Butler | **Peer Reviewer/s**: Felicity Brand, Kieran Morgan | **Expert Reviewer/s**: Saul Carliner

Breaking into the technical writing profession offers attractive benefits, such as lucrative pay and remote work options. Entering the competitive field of technical writing doesn't require a fixed educational path, making it an ideal second career for many. This chapter guides you through the journey from leveraging your existing skills to navigating job applications and interviews. Highlighting real-life stories and practical tips, this chapter is a must-read for anyone considering or transitioning into a career in technical writing.

Who Should Read This

- Aspiring Technical Writers
- Beginner Technical Writers
- Cross-Domain Professionals

CONTENTS

4.1 Introduction	52
4.2 Getting Qualified as a Technical Writer	54
4.3 Continuing Professional Development	57
4.4 The Application Process	57
4.5 The Interview Process	73

4.1 Introduction

With attractive salaries, great freelance rates, and benefits such as the ability to work fully or partly remotely, technical writing is a profession that many folks want to break into. Understandably, it's becoming ever more difficult to do so as the competition heats up and word gets out.

Technical writing isn't yet a mature profession in the way accounting, medicine, and law are. Those fields have a well-defined (but narrow) path from university through to professional practice. On the plus side, that means that technical writing can be a great second-career choice for folks who've built up their technical or writing expertise over the years in a different role.

As long as you can communicate effectively, have the aptitude and curiosity to learn new technologies, and can explain new concepts to others (especially through writing), you have the potential to be a technical writer. As a lifelong learner, you can become a great one!

Here are some of the diverse entry paths that writers have followed to get into technical writing:

4.1 Introduction

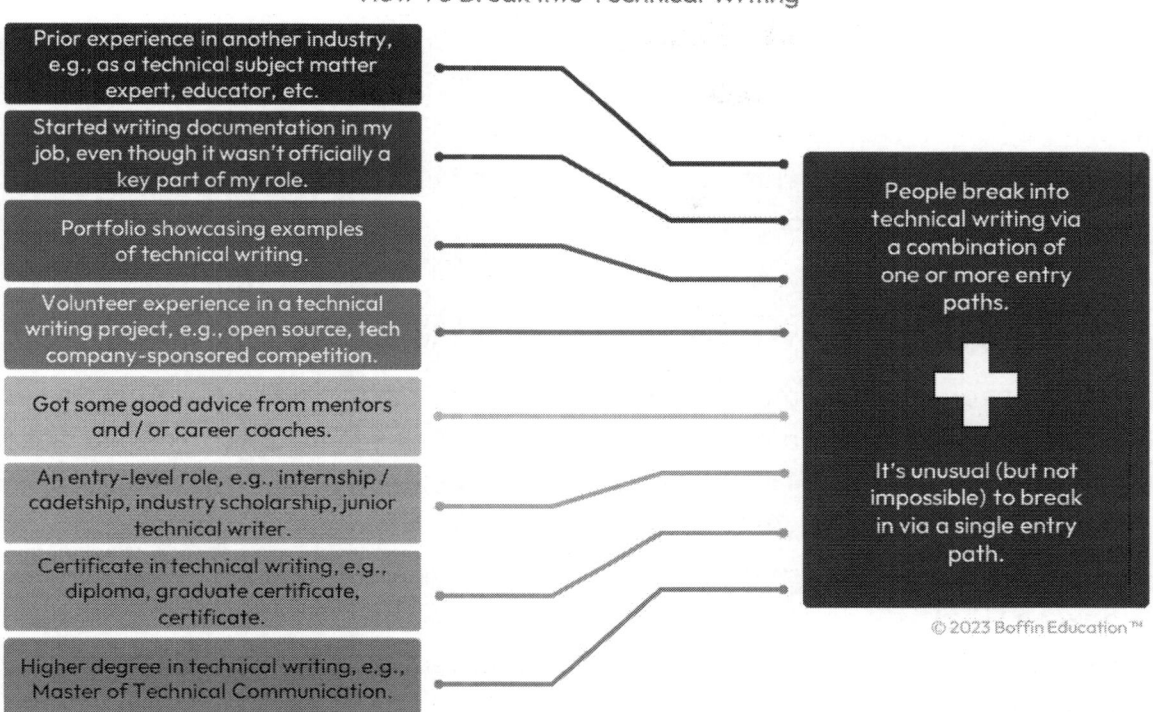

Figure 11: How to Break into Technical Writing

Breaking into the profession is the hardest part, but once you have a few years of experience under your belt and you've established a reputation for yourself, you'll never look back. Technical writing is a highly in-demand profession, and the more experience you have, the more employable you'll be.

Chapter 4 Breaking into Technical Writing

> **Voice of Practitioner**
>
> **Dina**
>
> - **Role**: Senior technical writer with 25+ years' experience in technical writing, technical documentation management, and information architecture
> - **Location**: Quebec, Canada
> - **Expertise**: Software, process and procedure, hardware documentation
>
> "You're not going to get hired unless you have really strong technical skills now. I was lucky to get hired without them back in the old days, but I don't think I'd get a job like that these days with the skills I had. To get your foot in the door, learn some basic skills that prove you're technical—even if you're not super technical! For example, how to document APIs, Postman, Open API spec, a little bit of Python or Java, C-sharp, and so on. The way the market is now, unless you have twenty years of experience behind you, you'll need those skills. This is what the recruiters are looking for and a lot of hiring managers in my field."

4.2 Getting Qualified as a Technical Writer

Unlike many professions, technical writing doesn't require a specific university degree. If you're not a member of a professional technical writing association (it's not mandatory), you won't need to get recertified by doing professional development courses that earn you credit points. However, unless you already have a very strong background in tech and writing, a technical writing qualification may be the single barrier to entry that your résumé is missing.

There are several benefits to gaining a technical writing certification:

- To signal to potential employers that you're **serious about a career in technical writing** because you've invested your own time and money gaining a qualification.

- To **formally recognize** your technical writing skillset. Even if you know you have the right skills, getting a qualification is an easy way to prove to potential employers that you have what it takes.
- To **broaden your skillset** so you can hit the ground running when you land your first technical writing gig. This will make your first years as a technical writer much less stressful.
- To **prepare yourself for the future** when you're an older tech writer—and the industry matures to the point where formal qualifications become the norm.
- To **establish your brand** as a qualified professional with credibility in the technical writing community and allow you to build your reputation as an expert, if that's your path.

4.2.1 How to Choose the Right Technical Writing Course

If you're considering investing in a technical writing course, there are a variety of options available to suit your budget and time commitments.

- **Universities and Colleges**: These institutions often offer courses that provide an extensive grounding in both the theory and practice of technical writing. These accredited programs are globally recognized but may require a significant investment of both time and money.
- **Professional Associations**: Courses offered by associations like the Society for Technical Communication (STC) are less time intensive and usually more affordably priced. However, they may not offer the depth of content found in university courses.
- **Private Education Providers**: Private providers, such as Boffin Education, offer highly practical courses to equip you with the practical skills and theoretical knowledge to become an expert. These courses range from affordable, self-paced micro-learning courses to hands-on, instructor-led sessions.

Each type of course has its own pros and cons, so weigh your options carefully. For more information on Boffin Education's technical writing courses—which are based on the same well-researched information as our books—visit our website at https://boffin.education/ today and take the first step toward advancing your skills!

4.2.2 Do I Need to Join a Professional Organization?

Unlike more mature professions, such as accounting, medicine, and law, it's not mandatory to be a member of a professional technical writing association to work as a technical writer. Nor do you need to pass an examination set by a professional association before you can practice. This may change one day, but for most technical writing roles, it's still optional at present.

So why join? Professional organizations provide job and networking opportunities, access to articles and training, and significant discounts on training, lectures, books, and conferences. Perhaps the most valuable part of membership is access to a network of professionals, which provides the opportunity for you to tap into their experience.

Here are some of the larger technical writing associations (check for local branches in your country):

- European Association for Technical Communication
- Society for Technical Communication
- IEEE Professional Communication Society

The world of technical writing has many lively online communities. These provide great opportunities for beginners to ask more experienced members for their advice on everything from technical questions about writing, to career advice. Here are some of the more popular ones:

- Write the Docs
- Reddit Technical Writing Community

- LinkedIn Groups, for example: Technical Writer Forum (45,000+ members), Documentation and Technical Writing Management (30,000+ members), and Technical Writing & Content Management (20,000+ members).

4.3 Continuing Professional Development

Technical writing is a rapidly evolving industry. That's partly what makes it such an exciting field. It also means that it's important to stay up-to-date with tools and technology in your industry so that you remain employable over the long term. You should devote time to continuing to develop your skills, both soft and hard. Consider it an investment in yourself. Writing in particular is a skill that can always be honed and improved, even after many years of practice.

Like many other jobs, opportunities in technical writing often arise from who you know, not what you know. Spend time building your network. Do this by joining technical writing meetups, attending conferences, and being active in online communities. If you're looking to move into management or move on to a role in a new organization, you can lean on your network for friendly advice.

4.4 The Application Process

To apply for a technical writing job—particularly if it's your first role—you'll need a few essentials, just like most other "white collar" jobs:

- A résumé
- A cover letter
- A LinkedIn profile

One thing that sets technical writing apart is the ability to showcase your excellent work. This is a great way to prove your skills—and employers often request it in an interview. So you'll need one other essential:

- A portfolio of your own writing samples

To help make this process less daunting for you, we've broken down each of these essentials below, giving you tips plus dos and don'ts to give you the best chance of landing a new role—and understanding what's expected of you.

4.4.1 Résumé

You must have a résumé or CV to apply for just about any technical writing job.

> **What Does That Mean?**
>
> **Résumé or Curriculum Vitae (CV)**
>
> A résumé (also known as a CV, or curriculum vitae) is a summary of your skills, responsibilities and achievements (work experience), and qualifications. Its purpose is to snag you an interview by proving to a prospective employer that you have what it takes to do the job. Think of it as a foot in the door to get you through to the interview stage.

Here's a list of dos and don'ts for writing your résumé.

Table 3: Dos and Don'ts for Technical Writer Résumés

Do	Don't
Be concise, and put the most relevant information on page 1. No hiring manager wants to wade through an essay-length résumé—in fact, many resumes aren't even read beyond the first page. However, expectations vary between countries. In the United States, the norm for résumés is generally one to two pages. However, in some jurisdictions (such as the UK, Canada, and Australia) a longer résumé or CV can be acceptable. Talk to a recruiter in your industry if you're unsure.	Don't include irrelevant experience. Leave trivial details off your résumé. Employers want to read about your technical writing accomplishments, not that you were responsible for minute-taking in your high school debating club. Additionally, if you have strong experience in something but you no longer want to go down that path, minimize or remove it from your résumé, or it will keep getting picked up in keyword searches by recruiters.
Be professional. Include a professional-sounding (or at the very least, inoffensive) email address, not the risqué or jokey one you had in high school or college.	Don't give a potential employer reason to discriminate (even unconsciously) based on irrelevant factors such as your: Marital status Religious affiliation Physical appearance (age, race, gender, disability) Although profile pictures are the norm on professional networking sites such as LinkedIn, you don't have to include them in your technical writing résumé.

Do	Don't
Be specific and focus on results when describing your accomplishments. Write "Wrote 11 successful grant applications" instead of "worked on grants." If you're coming from another industry, frame your transferable skills so that recruiters and hiring managers can easily understand how they apply to technical writing.	Don't lie or overstate your experience. If an organization needs a senior writer with specific skills that you don't have and you somehow land the role, you won't be successful or happy, and that will be bad for everyone—and most of all, your reputation.
Follow a standard résumé format. The most common formats are chronological or functional, which is sometimes used when switching careers. Whichever one you choose, be consistent and ensure it makes sense. A chronological résumé lists your experience in descending order of most to least recent. A functional résumé lists your experience under topical headings, such as "technical writing experience," "business analysis experience," and so on.	

Do	Don't
Follow good technical writing design principles when you're formatting your résumé. If you're using Microsoft Word, make sure you use consistent heading styles, and don't introduce any ad hoc formatting. For more information on styling documents, see Chapter 14: Design Stylesheet.	
Edit your résumé for typos and errors. Your résumé doesn't need to be a work of art, but it does need to be free of typos, grammatical errors, and other mistakes. Particularly if you're applying for technical writer roles, attention to detail is a key soft skill that employers will be looking for.	
Include keywords from job ads. Many résumés are scanned automatically by software that looks for keywords that match the job ad before they're ever seen by a human. So if the job requires a key credential, skill, or experience with a particular technology, make sure it's mentioned in your résumé—provided you do have that experience, of course.	

Do	Don't
Send it to a friend or family member to review. When you're crafting your résumé, you'll probably redo it so many times that you'll lose perspective on it. Having a fresh pair of eyes can find mistakes you missed, suggest better ways of phrasing your responsibilities and accomplishments, and give you a better idea of how it might land with employers.	

4.4.2 LinkedIn Profile

LinkedIn has become the default place to express your professional identity worldwide. Having a LinkedIn profile is essential for job seekers as it allows recruiters to find you using keyword searches and easily contact you. It also makes it easier to network with other technical writers and even apply for jobs using your LinkedIn profile (though it's a good idea to still have a résumé).

Here's a list of dos and don'ts to help you craft a strong LinkedIn profile.

Table 4: Dos and Don'ts for Technical Writer LinkedIn Profiles

Do	Don't
Add a photo of yourself. Unlike on a résumé, photos are the norm on LinkedIn. Add a professional-looking photo of yourself, cropped to show only your head and shoulders, against a simple or plain background. Make sure you're the only one visible in the photo.	Don't give a potential employer any reason to turn you down. This includes anything potentially contentious, such as liking a politician's LinkedIn post, even if you're not intentionally being partisan. Savvy employers will scour your LinkedIn profile and anything you've posted or liked to see if there's anything in there that might cause scandal or simply be against their personal beliefs.
Be mindful of posts you're tagged in, particularly where they're accompanied by pictures. Check your settings on your personal social media as well (Instagram, TikTok, YouTube). Are your personal photos visible to anyone? If so, do they portray you in a professional light? Even something as innocuous as enjoying a glass of wine might seem unprofessional to some employers if it clashes with their strongly held beliefs.	Don't post "jokey" or sarcastic responses to other people's LinkedIn posts, however tempting that might seem. Keep it professional, even if that means being a little bland. Just like anything politically controversial, it's best to keep it vanilla on LinkedIn. Save your snarkiest commentary for your alter ego on Reddit or somewhere a potential employer won't link it back to you and find offense.

Do	Don't
Add a concise headline that briefly describes what your key strengths are and how those relate to the career you're aspiring to. Here's an example; "Wordsmith with Technical Acumen	Self-Made Coder	Customer Service Professional	Aspiring Technical Communicator."	Don't be excessively negative in your posts. Some folks can't seem to help posting very negative comments, particularly about recruiters and the hiring process. Whatever your personal thoughts or experiences (and we're not saying that people haven't have bad experiences!), keep your diatribes for another forum. If potential employers see you posting negatively, they may think you have a chip on your shoulder that could express itself in a negative, morale-eroding attitude at work.
Tell a concise story about yourself in the about section. This will be an example of your writing style and abilities, so take your time to make it shine. Be authentic so you sound like yourself (first person, and not too "salesy" or "corporatey"). Something like: "I'm Lee Anne. For the past two years, I've been writing tech support content. I love learning from my amazing SMEs and am looking for a role that values collaboration over competition."	

Do	Don't
Complete all sections in your profile. LinkedIn seems to prioritize complete profiles over incomplete ones in search results. Add any skills, education, work experience, awards, languages, certifications, and licenses you may have.	
Engage with posts from companies you want to work for by liking them, share interesting articles about technical writing, join industry groups, and react to and comment on other people's posts. LinkedIn seems to rank active users more heavily in recruiter search results.	
Ask for recommendations from former managers, colleagues, or mentors. Even if you were in a completely different role from the one you're aiming for, having a glowing review of your communication skills on your profile will emphasize to a prospective employer that you have the right attitude to be a technical writer.	

> **Tip**
>
> **Customize Your LinkedIn URL for Your Résumé**
>
> You can customize your LinkedIn profile URL so it doesn't have a mess of letters and numbers at the end. This will look better when you include it in your résumé.

4.4.3 Cover Letter

When applying for a technical writing position, you'll often need to include a cover letter. In a few paragraphs, a cover letter gives you the opportunity to explain why you're interested in the role and the company, and the value you'll bring to the role through your unique combination of hard and soft skills, experience, and accomplishments.

In a writing-heavy field such as technical writing, a cover letter carries extra weight. It's a great opportunity to showcase your writing skills. On the flipside, poor writing or grammatical errors will reflect badly on you and be glaringly obvious to a potential employer. Because writing is a core part of the job, recruiters and hiring managers will scrutinize any writing you provide during the application process, so make sure it reflects your talent.

Table 5: Dos and Don'ts for Technical Writer Cover Letters

Do	Don't
Tailor your cover letter for the role. This goes for your résumé as well. Look at the sort of qualifications, skills, and experience the job ad is asking for, then see if you can find examples in your career where you've already done those sorts of things. Highlight these in your cover letter by summarizing them in bullet points.	Don't use generic copy-and-paste letters that are the same for every role. At the very least, make sure you address the hiring manager or recruiter by name, rather than "Dear Sir / Madam."
Keep cover letters to less than one page. They should be brief and to the point, highlighting only your most relevant skills and accomplishments.	
Try to put yourself in the shoes of the hiring manager or recruiter. The more directly and obviously your cover letter answers the question "How does the candidate meet my requirements for [key qualifications / skills / experience]," the easier it is to say yes to that candidate.	
Reference the job title and job ID (if there is one) in your cover letter. Remember that hiring managers and recruiters often recruit for multiple roles at once, and sometimes they lose track of things.	

Do	Don't
If time permits, ask friends or family members to review your cover letter—along with your résumé—before you send it to get a fresh perspective.	

 Tip

Don't Have Much Tech Writing Experience?

Cover letters are especially useful when you're new to the industry. If you're transferring into technical writing from another field, you can use your cover letter to explain how your previous experience and skills are relevant and transferable. If you're a recent graduate from a school, bootcamp, or certification course, you can use your cover letter to explain why you chose technical writing and what skills you can offer the organization.

4.4.4 Portfolio

Portfolios are excellent tools for showcasing your ability to clearly communicate technical concepts. In fact, many technical writing roles *require* at least several writing samples with your application. Portfolios can be especially valuable to aspiring technical writers, as they're an easy way of proving to a potential employer that you *can* do technical writing, even if you don't have much formal technical writing experience. For experienced writers, portfolios are an excellent way to showcase to potential employers your amazing writing and presentation skills, which their team will benefit from.

4.4 The Application Process

Portfolios can take many forms, varying wildly depending on the time and financial investment you're prepared to put into them. What matters is that your writing samples are high quality—well-written, structured, and formatted—demonstrate your understanding of technical writing concepts, and are easy for potential employers to access.

Technical writers often use a variety of different formats for their portfolios:

- A personal website with a custom URL
- A content management system such as WordPress
- A code hosting platform such as GitHub
- A document hosted online (such as a Google doc) that contains summaries of work samples or links to them

Be careful when you're preparing your portfolio that you don't disclose any sensitive intellectual property you don't have permission to share.

 What Does That Mean?

Portfolio

A portfolio is a collection of your own work samples from organizations you've worked for and that you have permission to showcase to prospective employers.

Good prompts for writing sample portfolio pieces include:

- How to write an email (use the device/program/browser of your choice)
- How to download images from a device
- How to download an app to your phone and use it
- How to order groceries online
- How to send images through a social media platform

- Compare and contrast the use of a laptop/desktop with that of a tablet/phone to complete a task
- Compare and contrast the use of two different programs to complete the same task
- A troubleshooting guide for something in your home/daily life that sometimes malfunctions
- Table or graphic comparing two programs/products
- Define terms (app, program, HTML, CSS)

For more advice about choosing writing samples, we recommend you read Sue Arkin's post *The Dos and Don'ts of Writing Samples*[10] in her blog, <u>Tested Writing</u>.

> **Tip**
>
> **How Do I Build a Portfolio if I've Never Done Any Technical Writing?**
>
> If you're an aspiring technical writer and don't have any published samples from a previous job, that's okay. Many aspiring technical writers find themselves in the same position. Consider creating some documentation using the principles outlined in this guide. If your samples showcase your technical skills as well—for example, some amateur coding projects you've done—even better.

Table 6: Dos and Don'ts for Technical Writer Portfolios

Do	Don't
Writing samples should be long enough to provide context and showcase your communication and or technical abilities but short enough that a hiring team can review them easily. Keep each sample to less than a page. If possible, use pieces that are most relevant to the role you're applying to.	Avoid weak samples that don't showcase your best abilities, or samples that aren't tech-docs specific. For example, some folks include a broad range of writing projects, such as short stories, screenplays, poetry, and so on. This is best left for your personal website rather than your professional portfolio.
Include sufficient context alongside your samples so that employers can understand them: the organization (or industry if you're not allowed to mention the organization by name), audience, tools or techniques you used to write the samples, any challenges you had to overcome, the positive impact your pieces had (e.g., reduced support calls, or were described in glowing terms by a customer in an online forum).	Avoid highly idiosyncratic portfolio sites that may make you seem just a little too quirky. Once you're in the workforce and your teammates get to know you, you can judge whether it's safe to let your hair down and show your true self. It's all too easy for a potential employer to turn you down based on an impression from a website before they get a chance to know you—and appreciate your unique strengths.
Include a link to your portfolio in your LinkedIn profile and at the top of your résumé.	
Try and make your portfolio links public if you can. It's no good having a portfolio if a recruiter can't see your work!	

Chapter 4 Breaking into Technical Writing

Do	Don't
Ensure that you have permission to share any proprietary work that you have done for a former employer—and that it isn't behind a paywall or nondisclosure clause in your contract.	

 Voice of Practitioner

Jerome

- **Role**: Technical writer with 4-+ years' experience and 10+ years' language and writing tutoring
- **Location**: Washington, United States
- **Expertise**: Software documentation

"My portfolio has been one of the strongest things to qualify me for roles. If you don't have any experience in tech writing, it is entirely up to your portfolio. People didn't even look at my résumé until I had a strong portfolio. To get my first tech writing role, I volunteered for open-source work and landed a gig at the Google Season of Docs project. I worked with the Madplotlib Project,[11] and they enjoyed my proposal. I developed a lean style guide, some docs, and documented an entry path for someone who was entirely new to programming, noting there were loads of online resources such as YouTube available. I noticed gaps that needed to be in the entry path that weren't in the docs and highlighted that to them. I like to say 'open source opens doors.' However, open-source projects are developers doing it out of passion, it's a labor of love, so there can be a high barrier to entry there. It's a tight-knit, closed community."

4.5 The Interview Process

Many folks dread interviews. It may surprise you to learn that most hiring managers don't exactly love them either. From the hiring manager's point of view, interviews can be draining ordeals that require high levels of concentration. They may be conducting many interviews in a single day—and not just for a single role. Managers know they've got an hour or less to understand how well you can write, empathize with your audience, work well with others, and stay consistent in your editing. As a candidate, they're a great opportunity for you to showcase not just your accomplishments and technical skills, but also your soft skills.

So how does it work? Technical writer interviews usually consist of at least one interview (phone, video, or in person) and often include a writing test as well. A short, more informal interview process, which may happen if you've been personally recommended by a trusted colleague, usually involves a single interview with the hiring manager.

A longer, more formal interview process may involve an initial interview with a recruiter, an interview with the hiring manager or a panel of multiple interviewers, and a chance to meet the rest of the team you'd be working with. This also gives them a chance to meet you and provide feedback to your manager on whether they think you're a good fit for the team.

Interviews for technical writers focus on both soft skills and hard skills. Common questions include:

- What does a successful technical document look like?
- How would you keep your audience engaged?
- Imagine you're about to document a new product or feature. What does that process look like?
- Tell me about a time you handled conflict in a school, work, or other group setting.
- What is the purpose of a style guide?

Following a successful interview process and a job offer, you and the organization will have to agree on employment terms. Salaries can vary widely depending on the industry, your technical skills, and your level of experience. If you're a newbie, you may not have much bargaining power. Location also matters. For example, even within the United States, the cost of labor in Silicon Valley is not the cost of labor in Kansas City. For more information about salary, see Write the Docs' Salary Surveys[12] on their website.

 Tip

Build Confidence by Rehearsing Your Answers

It's very easy to get flustered during an interview, particularly if the interviewer throws a curveball question at you that you weren't expecting. The last thing you want to do is dissolve into a puddle of "uhms" and "ahhs" in the middle of a session. To avoid this, grab a list of potential interview questions and rehearse your answers to them until you're confident that you can respond to them. You don't need to recite the answers verbatim in the interview, so repeat them just enough times that the key points readily spring to your mind.

4.5 The Interview Process

 Voice of Practitioner

Dina

- **Role**: Senior technical writer with 25+ years' experience in technical writing, technical documentation management, and information architecture
- **Location**: Quebec, Canada
- **Expertise**: Software, process and procedure, hardware documentation

"What I look for when I interview someone for a tech writing role is extreme curiosity: someone who's not going to give up when the SME says, 'That's not important.' I'm looking for someone who truly cares. I've hired support people because I've seen the amazing documentation they've written for themselves. They've used no tools, but they have this desire to figure out a solution to save the user some time. You can teach a person who's smart and motivated and willing to learn anything, but you have to have this thing in your head that won't rest until you have an answer to your problem. It's the people who stay up at night trying to solve a problem that I hire in a heartbeat because they're always going to be improving, always getting better. That's an innate skill that you can't teach. You can be great technically, but if you don't care figuring out how to get something, you're always going to be 'just ok.'"

4.5.1 Writing Tests During the Interview Process

So, you've gotten an interview for that coveted technical writing role…and they've told you that you'll have to do a test. Don't panic! Try and think of the writing test as an opportunity for you to showcase your excellent skills and in doing so stand out from the competition, rather than as an ordeal to be endured.

It helps to understand why hiring managers and recruiters use writing tests. Here are the key reasons:

- A writing test shows how you actually work under pressure vs. your writing samples, which showcase your best work that you've had the opportunity to finesse at your leisure. The writing test will demonstrate the typical volume and quality of work you can be expected to produce within a certain time.

- It tends to weed out the pretenders pretty quickly. These are the folks that lie on their résumé about having extensive experience or skills that they don't—or those who didn't create their own portfolio samples themselves.

- Finally, writing tests show how well you can follow instructions. Your potential employers will be assessing not just your communication ability, but also how well you listened to them. The ability to listen to stakeholders (including your boss) and pay close attention is an absolutely essential technical writer skill.

A typical writing test takes between one and two hours and tests your writing and editing skills. The hiring manager is looking for consistent, error-free work that demonstrates an understanding of technical writing principles.

Chapter 5 Career Growth and Survival

Lead Writer: Amanda Butler | **Peer Reviewer/s**: Felicity Brand, Kieran Morgan | **Expert Reviewer/s**: Saul Carliner

This chapter is a go-to resource for technical writers at all career stages. It offers guidance on everything from overcoming early-career jitters to making pivotal career decisions. It outlines the various seniority levels and the skills required for each as well as giving an in-depth look at managerial roles. Moreover, it shows how technical writing can be a launchpad to other career opportunities, making it a must-read for anyone seeking to navigate and excel in the field.

 Who Should Read This

- Aspiring Technical Writers
- Beginner Technical Writers
- Career Advancers
- Managers of Technical Writers
- Cross-Domain Professionals

CONTENTS

5.1 Introduction	78
5.2 Transitioning to Your First Technical Writing Role	78
5.3 Career Paths within the Industry	83
5.4 Careers after Technical Writing	90

5.1 Introduction

Embarking on a career as a technical writer marks the beginning of a journey filled with opportunities for growth and development. This chapter guides you through the various stages of your career, from the initial phase of settling into your new role to exploring advanced career paths within the technical writing industry.

We begin by addressing the common feelings of imposter syndrome among new technical writers, providing insights and advice from seasoned professionals across different countries and industries. Their experiences will be especially helpful if you're just starting out in the field.

As you progress in your career, you'll encounter decisions about your path forward. This chapter discusses the choice between pursuing an individual contributor (IC) track, where you hone your craft independently, and the management track, where you lead and mentor others. We outline the typical levels of seniority in technical writing, from junior to managerial roles, and examine the challenges and rewards at each stage.

Understanding the possibilities inherent in these career tracks is essential, whether you're a junior writer learning the ropes or a senior writer influencing key decisions. By the end of this chapter, you will have a clearer understanding of how to navigate your career, capitalize on your strengths, and make informed decisions that align with your aspirations.

5.2 Transitioning to Your First Technical Writing Role

Having outlined the broader landscape of career progression in technical writing, let's focus on the very beginning of this journey: your first role as a technical writer. Stepping into the professional world, especially in a field as ever-evolving as technical writing, can be both exhilarating and daunting.

5.2.1 Surviving—and Thriving!—in Your First Technical Writing Role

Okay, so you've gotten your first job as a beginner technical writer. Congratulations! Are you feeling impostor syndrome yet? Don't worry, that's normal. We've all felt that at some stage in our careers. To support you in this initial phase, we've compiled practical tips from experienced technical writers who have navigated this path successfully. Their shared wisdom, drawn from various stages of their careers and diverse professional backgrounds, will provide you with valuable insights as you embark on your own journey.

 Voice of Practitioner

Layale

- **Role**: Junior technical writer with 3+ years' experience
- **Location**: Quebec, Canada
- **Expertise**: Process and procedure documentation

"Get a book and start reading! If you're starting off as a tech writer, before you write anything, read a book that will inform you about what you will need to know in your technical writing process."

Voice of Practitioner

Carly

- **Role**: Technical writer with 3 weeks' experience and 18 months as a knowledge management specialist
- **Location**: Queensland, Australia
- **Expertise**: Process and procedure documentation

"Get a qualification. Do some formal training before you start. I wish I'd had a better understanding of industry standards and 'what good looks like.' I think my experience was specific to my team and my industry."

 Voice of Practitioner

Anh

- **Role**: Technical writing intern with 1 month's experience
- **Location**: Pennsylvania, United States
- **Expertise**: Software documentation

"Don't be afraid to ask questions and reach out to more experienced people. I know a lot of people are like me; they're afraid to bother others, afraid they'll waste someone's time. But it's natural for people to help newcomers, as many received significant help from others early in their careers."

5.2 Transitioning to Your First Technical Writing Role

 Voice of Practitioner

Nellie

- **Role**: Documentation specialist with 6 months' experience
- **Location**: Pennsylvania, United States
- **Expertise**: Software documentation

"When you're confused, be present, and let the confusion wash over you. You don't have to make sense of it all at once; it eventually will. Go take a walk or whatever, and let diffuse thinking happen—your brain will make sense of it later."

 Voice of Practitioner

Amanda

- **Role**: Technical documentation manager with 10 years' experience
- **Location**: Texas, United States
- **Expertise**: Software documentation

"Ask all the questions. Don't try to act like you know the answer when you don't. I know culture varies by company and country, but engineers and other SMEs tend to respect people when they admit to not knowing. They'll eat you alive if you pretend to know and you don't."

Voice of Practitioner

Colin

- **Role**: Senior technical writer with 5 years of formal technical writing experience and 7 years of experience creating documentation under different job titles
- **Location**: Quebec, Canada
- **Expertise**: Software, hardware, process, and procedure documentation

"Find a mentor. There are a lot of very talented people in this space who love talking about it. Groups such as Write the Docs are filled with individuals who are very happy to share their experience and tips. It's very important to find a mentor at the beginning of your career to encourage and support you because it can be very difficult at the start."

 Voice of Practitioner

Sarah

- **Role**: Senior technical writer with 5 years' experience
- **Location**: Toronto, Canada
- **Expertise**: Software documentation

"Keep a record of everything you're doing so you can bring it up during your performance reviews—what you did, how you did it, and the outcome. I didn't do that in my first couple of positions and regretted it. Keep records of your accomplishments—you'll thank 'past you' for doing that."

Insight

Insider Perspectives: Technical Writer Interviews

When we wrote this book, we interviewed more than twenty technical writers at all stages of their careers—from interns with only a month's experience to senior writers with over thirty-five years' experience reflecting on long and successful careers. We talked with documentation managers responsible for leading teams across continents and aspiring technical writers wishing to break into the industry. We asked questions about every aspect of their careers—from the tools they use to their advice for newcomers to the profession and what they think the future holds. You'll see snippets from these interviews used throughout the book, which we've collected in Interviews.

5.3 Career Paths within the Industry

After you've successfully applied to a job, gotten hired, and made it through your first year, what's next? Over time, you'll learn new skills and take on new responsibilities. As you progress in your career, you'll eventually need to choose between two career tracks: the individual contributor (IC) track or the management track, or possibly a combination of both.

We've identified four fundamental levels of seniority in technical writing: junior (entry-level), technical writer (intermediate), senior (advanced), and manager (people leader).

Chapter 5 Career Growth and Survival

Levels of Seniority in Technical Writing Roles

Ayo
Junior Technical Writer

"I'm a junior technical writer—but I feel like I have a technical writer's job! I have two years' experience. I have a bachelor degree with a minor in information technology, and I've done two online technical writing certification courses."

Francis
Technical Writer

"I have four years of technical writing experience. I'm currently studying my bachelor of science, engineering and technology. I'm constantly trying to build my skill set. My aim is to become a technical and product expert in my company so I can take the next step up in my career. It doesn't have to be a technical writing role."

Dina
Senior Technical Writer

"I have more than twenty-five years' experience in technical writing and information architecture. I have a Master of Literature and I'm currently finishing a Master of Information Science. I like to connect people so problems can be solved and things made better. That's my dopamine hit. I'll often stop and ask junior colleagues if they need help with anything."

Amanda
Technical Documentation Manager

"I have over ten years' experience in technical writing and tech docs management. I have two degrees; one in physics and another in journalism. My passion is finding out where people want to grow and what their strengths are and plugging them into opportunities. It's like playing Tetris with people! Am I allowed to say that?!"

© 2023 Boffin Education™

Figure 12: Levels of Seniority in Technical Writing Roles

In different organizations, countries and industries, these roles may go by different names. Some organizations with extensive and mature technical writing departments might have numerous grades and role classifications, and some smaller organizations might simply have one role: technical writer.

We've included some general guidelines below as to the typical qualifications and experience for entering these roles. Remember, there's no hard-and-fast rule. Don't think you can't be a senior technical writer or manager simply because you don't have many years of experience.

Also, don't be worried if you do have many years of experience and you're not a senior writer. Many seasoned technical writers simply prefer to focus on producing great work without the additional mentorship and people leadership roles that a senior job title carries.

5.3 Career Paths within the Industry

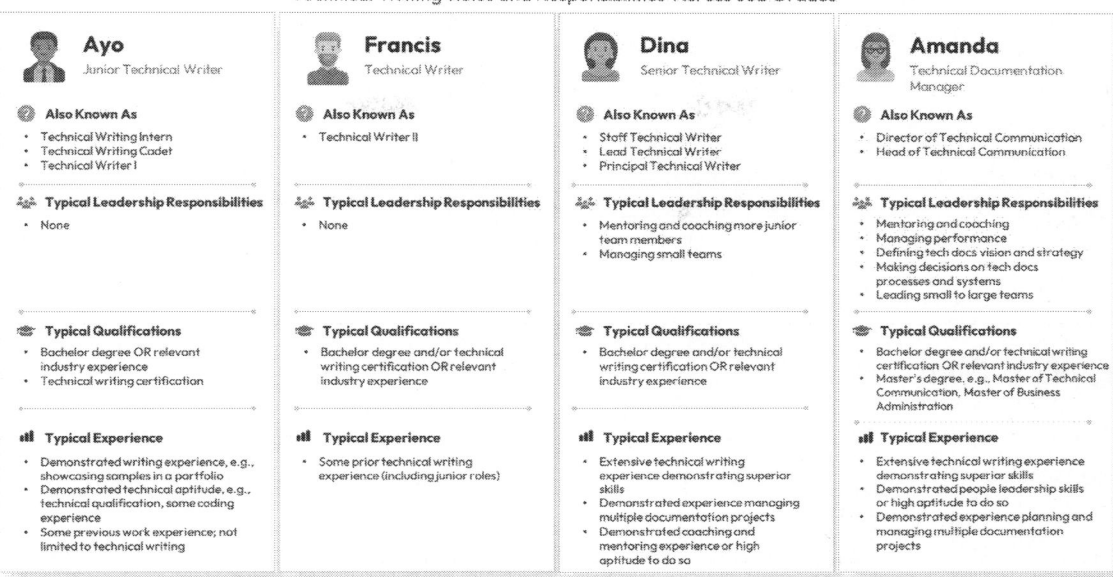

Figure 13: Technical Writing Roles and Responsibilities Across Job Grades

 Voice of Practitioner

Amanda

- **Role**: Technical documentation manager with 10 years' experience
- **Location**: Texas, United States

"These days I see a trend to avoid 'minimum qualifications,' 'years of experience,' or anything that could gatekeep candidates from roles due to privilege or lack thereof. We're trying to get people in who can demonstrate that they have the skills they need to do the job rather than a specific qualification, like through coding bootcamps."

5.3.1 The Individual Contributor (IC) Track

Many (if not most) technical writers are individual contributors (ICs). This simply means they don't manage other people, which suits a lot of folks just fine, even those who have many years of experience.

 What Does That Mean?

Individual Contributor (IC)

Individual contributor (IC) is industry jargon for someone who doesn't manage other people.

The number of levels, or grades, in the individual contributor career path usually depends on the size of the organization. Here's an example of a typical IC career path:

1. Junior technical writer
2. Technical writer
3. Senior technical writer

Large organizations may have additional roles beyond senior: staff, principal, or lead, for instance. It's worth your while to check that you understand the seniority of the role correctly.

 Note

Job Titles at Small Organizations

Start-ups and small organizations often have a single technical writing role. In that case, your title may stay "Technical Writer" throughout your time at the company, even if your role grows and you take on additional responsibilities.

5.3.1.1 Junior Technical Writer Roles

As a junior technical writer, you should focus on developing your skillset through experience. It's okay to make mistakes! Your manager and more senior technical writers should be there to review your work before it gets published and give you feedback on how you can improve.

What else should you expect as a junior technical writer?

- You should receive tasks from your manager that are manageable within your developing skillset, such as for a single product or component, or a simple business process.
- Your manager or a senior technical writer should familiarize you with your organization's documentation process, systems and repository, style guide, and templates.
- Your manager should introduce you to subject matter experts, give you clear guidance for tasks, set expectations around the document review process and timelines, and make themselves or a senior technical writer available to answer your questions.
- You will contribute to (and may even own) a section of the documentation or a small set of relatively straightforward documents.
- As you move through the levels, you'll learn to work more autonomously. You'll eventually be able to make informed suggestions about how to fix bugs in software products, tweak features to improve their usability, or improve business processes.

The difference between a junior writer and a senior writer is not how well you write or even how long you've been in the industry. The difference is how independent you are, the depth of your expertise and product or organizational knowledge, and how much you influence others in the course of your work.

5.3.1.2 Senior Technical Writer Roles

As a senior technical writer, you will be expected to be a master of your craft with a deep understanding not just of technical writing principles, but of the technology of your industry and your organization's unique products and processes.

What else should you expect as a senior technical writer?

- You should work with your manager and subject matter experts to set your own tasks and priorities, negotiate with them to set realistic deadlines, and proactively let folks know if things change, for example, if it's looking like that deadline you set is now out of reach.

- You should be comfortable making recommendations to improve technical writing processes, the company style guide, definitions of terminology within the organization's dictionary, and anything you think may negatively affect the user experience in products you're documenting.

- You will use your knowledge to mentor and coach junior writers, giving them feedback and peer reviews on their work before it's published, answering their questions about their craft, guiding them so they understand what's expected, and upskilling them on the organization's systems, products, and processes.

As a senior writer, you will have the opportunity to "choose your own adventure"—even beyond technical writing. You may choose to go deep and learn everything about a product or technology and become an expert yourself, opening up opportunities to advance your career in fields such as product management. Or you may "go guru" and build your reputation and qualifications as an expert technical writer—perhaps even embracing a role such as an educator, passing on to others your advanced mastery of the craft.

For other examples of career paths, see [Career Ladders for Documentation](https://career-ladders.dev/) on https://career-ladders.dev/, and another example is the [Technical Writer](#) and [Technical Writing Management](#) levels in https://handbook.gitlab.com/.

5.3.2 The Management Track

Some technical writers manage teams of writers. These folks are usually called technical documentation managers, though they can also be called documentation leads or lead technical writers. Even more senior technical writing roles exist, where you will be expected to manage other managers, such as technical communications director and head of technical communication, all the way up to vice president, technical publications.

Don't think of the management track as a promotion over the IC track or a recognition of superior technical writing skills. It's a distinct career path with its own set of soft and hard skills that combine elements of industry and organizational knowledge, people leadership, strategy, project management, and process management. In fact, some people who aren't the strongest writers might be terrific managers—the two are very different skillsets.

As a manager of other writers, you'll most likely find you'll do less writing, but it will help your credibility and make it easier to empathize with your technical writing team if you set aside time to do *some* writing.

That being said, expect to focus most of your time on developing your leadership skills. Being a great leader means giving your writers the support they need to consistently produce high-quality documentation in a timely manner and building your team's processes so that turning out excellent documentation and meeting deadlines is as frictionless as possible for your team.

What else should you expect as a technical documentation manager?

- Assign tasks to writers of all seniority levels on the team and help writers prioritize their tasks.
- Track progress to ensure everyone is on schedule and any delays are promptly communicated.
- Provide different levels of supervision depending on the complexity of the assignment.

- Set expectations about each person's role on the team, responsibilities, and deliverables.
- Help remove obstacles that are blocking the writing team from creating good documentation.
- Resolve interpersonal issues between the writers, or between writers and their stakeholders.
- Assess the performance of team members, recognize and celebrate achievements, and address performance issues.
- Interview and hire new technical writers.

5.4 Careers after Technical Writing

Few folks spend their entire career at the same organization. Fortunately, the core soft and hard skills of technical writing are very transportable—that is, they're similar regardless of the organization or industry. As you gain experience in your craft, you should find it relatively easy to move into technical writing roles in other organizations or industries.

Technical writing can be a fantastic springboard into other careers. Technical writers who have worked in an organization or industry for a while often build up extensive specialized knowledge about products, processes, industry-specific laws and regulations, and so on. This wealth of knowledge can make it a relatively simple affair for technical writers with a good reputation to move into roles such as the following:

- Design or user experience (UX)
- Product management
- Project management
- Marketing
- Quality assurance (QA)
- Development, engineering, or research and development (R&D)

Voice of Practitioner

Francis

- **Role**: Technical writer with 4 years' experience
- **Location**: Manchester, United Kingdom
- **Expertise**: Software documentation

"Try and get expert knowledge in the product. It doesn't matter how great you are in AsciiDoc or Git, you're not measured to that; you're measured to how well you know the product. When you need quick turnarounds, or if you're working with someone who doesn't have time for you, if you've spent your spare time understanding the product you become a pretty safe person in the team and very valuable in the business. You also open up opportunities for yourself in the business to move into other roles if you want."

Chapter 6 Career Flexibility

Lead Writer: Amanda Butler | **Peer Reviewer/s**: Felicity Brand, Kieran Morgan | **Expert Reviewer/s**: Saul Carliner

This chapter provides a comprehensive look at flexible work options for technical writers: remote, cross-time-zone, and freelance work. Each mode comes with its own set of benefits and challenges. They offer opportunities for high autonomy and flexibility but also require careful consideration of factors like job security and team collaboration. It's a guide to making well-informed, rewarding career choices in the evolving landscape of technical writing.

 Who Should Read This

- Aspiring Technical Writers
- Beginner Technical Writers
- Career Advancers

CONTENTS

6.1 Introduction	94
6.2 Working across Multiple Time Zones	95
6.3 Freelance Work	97

6.1 Introduction

Technical writing is a great opportunity for those who enjoy remote or hybrid work—that is, working some days in the office and some days from home. Many technical writers do a hundred percent of their research, writing, and editing from their home office. All you need is a quiet location where you can concentrate, permission from your manager, and a stable internet connection that allows for videoconferencing and screen sharing.

It's important to note that remote work isn't for everyone:

- If you're a beginner, it might help your career to be in the office until you've learned the ropes. Being physically present might help you interact with others including your team and your subject matter experts (assuming they're also physically present in the office!) In doing so, you may soak up the office culture and norms more quickly than at home.

- If you're a technical writer who writes about hardware and you need access to it to make sure your documentation is accurate, then remote work might not be for you. You may need to spend at least a day or two a week on site physically learning how to use the hardware. Many hardware tech writers relish this aspect of their job.

- You may be required to be on site if you work for the government or any organization that requires you to work on confidential documents from a secure location.

- Finally, some organizations have very strict rules regarding remote and hybrid work. Some don't allow it at all, while others swing completely the opposite way, building remote work into their employee value proposition to help them recruit staff. Check what their expectations are before you start working for them and find out that their working arrangements aren't going to suit you.

Voice of Practitioner

Francis

- **Role**: Technical writer
- **Location**: Manchester, United Kingdom

"If you're working remotely, take care of your persona online. For example, if you're going to have a beard, trim it. If you wear a t-shirt, make sure it's not one the cat slept on. If you're on camera, pay attention when people are talking, especially your boss. Don't be typing when people are speaking, or looking like you're falling asleep, or not looking into the camera. This is a big thing in the workforce. It doesn't matter how passionate you are about the company, they'll always pick the people who seem to have more of a positive persona for the opportunities."

6.2 Working across Multiple Time Zones

Working across different time zones is quite common, particularly for writers who work for large, global organizations. It can be challenging if your subject matter experts or fellow writers are spread across multiple time zones. You might find that you'll be expected to attend very early or very late meetings (or at worst, a combination of both) well outside your nine a.m. to five p.m. work window to accommodate colleagues in different regions. For this reason, it's best to check the locations and time zones of the people you'd be working with before you start your new job, plus the timing of any regular meetings you'll be expected to attend.

Insight

Working Remotely Can Be Isolating for Those in Different Time Zones

If you're the only member of the team working remotely and everyone else is co-located, it can feel lonely, and it's easy for the team to forget to include you. If you find yourself in this situation, be assertive in reminding the team about the need to include you in all conversations and update you about any ad hoc conversations that may have happened in the break room or outside scheduled meetings.

 Voice of Practitioner

Amanda

- **Role**: Documentation manager
- **Location**: Texas, USA

"I work with people across the globe. My boss is in England. Normally I wake up to a ton of Slack messages. A lot happens while I sleep! The first thirty minutes of my day are spent catching up on what's happened overnight."

Voice of Practitioner

Robert

- **Role**: Technical documentation manager
- **Location**: Massachusetts, USA

"My team is mostly split between North America and Europe, so my day is usually split in two chunks: a morning chunk where I talk to Europe, then in the afternoon I start to work with the interstate US folks. That's been working pretty well."

6.3 Freelance Work

Technical writing offers great opportunities for freelance work. Freelancing (also known as contracting) can provide lucrative hourly or daily rates well beyond what a salaried role can offer, plus the flexibility to work your own hours in the time and place of your choice.

For technical writers with good experience under their belt and a great professional network, it can be extremely rewarding. However, it's not for everyone. Beginners without much experience may struggle to negotiate good freelance terms or get enough work. Many folks simply prefer the job security and career advancement opportunities that come with a salaried role.

Before going down the freelance path, you should be fully aware of the pros and cons. Here are some of them from our experience.

Chapter 6 Career Flexibility

Table 7: Pros and Cons of Freelance Work for Technical Writers

Pros	Cons
Enjoy premium rates well in excess of salaried roles—within limits. Many companies have freelance rates (or a range) defined for certain professions, and they won't go very far beyond this even if you're the Stephen King of tech writers. Savvy freelancers should develop good relationships with recruiters and use that relationship to find out what the industry norm is for rates, as well as specific ranges at organizations that may be willing to pay a premium.	Job security can be lacking for freelancers. There may be plenty of work in economic boom times and excellent freelance rates available. This can dry up very quickly in challenging economic times as organizations prepare to endure recessions. In such times, shedding freelance workers can be an expedient way for organizations to save money.
Set your own schedule. Freelancers who do piecework (short contracts that are often costed and quoted in advance) can complete work in their own time and juggle multiple clients simultaneously. This gives them the flexibility to focus on other things in life, like taking care of children and dependents, or focusing on passion projects.	Lack of career advancement and training opportunities. Usually, freelancers will be expected to invest in their own education and won't have access to training courses that full-time staff do. Organizations are often reluctant to invest in career development opportunities for freelance staff, particularly at a management level. They may even be prohibited by law from providing these opportunities in some countries.

6.3 Freelance Work

Pros	Cons
Set your own conditions, such as choosing the clients you'd like to work with, turning down projects that don't suit you, and potentially working from anywhere in the world (this is known as a "digital nomad"). Note: Although the digital nomad life may sound very appealing (and cost-effective), make sure you fully research the pros and cons before doing so. You may struggle to get paid if you don't have a bank account in your employer's country. Some countries impose stringent visa and tax restrictions on digital nomads, and your employer may frown on you working from overseas.	Freelancing adds additional (often hidden) costs and complexity to your finances. You should take these into account when you're comparing them to a salaried role. For example, you may need to consider: • Paying your own tax to the government, which may need to be done regularly, maybe quarterly. • Obtaining your own health, life, and income protection insurance. • Contributing toward your own retirement fund. • Inconsistent cash flow throughout the year, including if you take a significant chunk of time off for a vacation. • Challenges when you try to rent or buy a home or car. Landlords and banks often require evidence of consistent income and may prefer not to lend to freelancers.

Pros	Cons
	If all of this sounds daunting, talk to a freelancer in your industry to see how they manage it. Complexity can be minimized by investing a small percentage (maybe 3 percent) of your income with an intermediary such as a payroll service or accountant who will manage compulsory taxes, insurance, and retirement fund contributions on your behalf.
Focus on producing work rather than administrative tasks such as attending team or departmental meetings, going through lengthy onboarding processes or mandatory training, participating in performance development cycles, and so on. However, you will need to invest time in your own administrative tasks, such as invoicing your clients, pitching for work, and bookkeeping.	You might feel like an outsider in the team due to your freelance status. Permanent members of staff may consider you a temporary member of the team, not worth investing the time establishing a relationship with. Some companies even issue different-colored ID badges or lanyards to freelancers so they can be easily distinguished from full-time staff. You'll definitely need to have the knack of establishing rapport with your SMEs very quickly and the skill of finding information by whatever means necessary in the shortest possible time.

Pros	Cons
Gain valuable industry experience when full-time roles are scarce, which might eventually lead to full-time roles, if that's your aim.	Most new writers benefit from working with and learning from experienced technical writers. When you freelance, you may find yourself working solo. You might find you have less opportunity to work with other writers, which can put you at a disadvantage when it comes to growing your skills. If you're already highly experienced, this might not matter so much to you. Having a strong network is essential for successful freelancers who are always on the lookout for the next contract.

Part 2 Process

Exploring the customization of the Technical Writing Process to align with unique project demands and organizational cultures.

Chapter 7 Tailor the Process

Lead Writer: Kieran Morgan | **Peer Reviewer/s**: Steve Moss

This chapter focuses on the customization of the Technical Writing Process to suit specific project needs and organizational contexts. It outlines a practical, iterative approach to adapting this process, one that emphasizes the importance of understanding an organization's unique requirements and modifies the process accordingly. The chapter also includes case studies demonstrating the application of this tailored process in different organizational settings and providing practical examples of its flexibility and adaptability.

Who Should Read This

- Aspiring Technical Writers
- Beginner Technical Writers
- Career Advancers
- Managers of Technical Writers
- Educators of Technical Writers
- Cross-Domain Professionals
- Consultants

CONTENTS

7.1 Introduction	107
7.2 [Theory] The Technical Writing Process	107
7.3 [Theory] Comparing Information-Development Frameworks	113
7.4 [Practice] Customize the Process	115
7.5 [Case Study] Case Study 1: Software Start-Up Company	118
7.6 [Case Study] Case Study 2: Multinational Engineered Products Company	121

Chapter 7 Tailor the Process

PROCESS

Inputs	Process > Tailor the Process		Outputs
• Collected Data, Information, and Knowledge • Stakeholder Interviews	**Theory**	• The Technical Writing Process • Comparing Information-Development Frameworks	• Customized Technical Writing Process
	Practice	• Download and Learn • Analyze Your Organization's Requirements • Add, Amend, Delete • Iterate and Improve	
	Samples	• Case Study 1: Software Start-Up Company • Case Study 2: Multinational Engineered Products Company	
	Templates	• Technical Writing Process Template • Technical Writing Process Checklist	

7.1 Introduction

If you scan this book's table of contents, you'll notice how it's structured around eight core phases: Plan, Design, Write, Edit, Review, Translate, Publish, and Manage. These represent the backbone of the Technical Writing Process, the macro steps that writers such as yourself carry out when crafting technical documentation.

But it's more than that. The process is a learning tool that weaves in concepts relevant to the practice in each chapter. Think of the Technical Writing Process as your roadmap to your craft. You can use it as a learning tool to build your knowledge of the fundamentals of our profession or as a practical tool to apply your knowledge in the real world.

In this chapter, we'll take a practitioner's view. We'll give you a bird's-eye perspective of this practical process, why it's so important, and how to customize it to suit your writing projects.

7.2 [Theory] The Technical Writing Process

The Technical Writing Process defines the eight core steps in technical writing projects: large or small, complex or straightforward. The steps aren't strictly sequential; in practice, there are plenty of loop-backs and decision points that lead to alternate routes through the process. Much of this iterative nature occurs in the write-review-update cycle, which we discuss in Chapter 17: Write Draft | [Theory] The Write-Review-Update Cycle.

So why present a sequential view if it's not strictly true? First, every project is slightly different—the high-level view of the Technical Writing Process encompasses the flexibility to accommodate this. Second, the simplified view helps you understand the big picture before you develop a deeper understanding of the myriad ways it can be applied in practice. All in all, it's a much easier way to comprehend the vast breadth of technical writing—and internalize the process as your mental roadmap—before you immerse yourself in the possibilities and nuances of application.

You'll notice that the process is divided into several sections:

- **Phases**: These are the macro steps of the Technical Writing Process under which everything is organized: Plan, Design, Write, Edit, Review, Translate, Publish, and Manage.
- **Activities**: These are the high-level steps that support each phase. In this book, roughly speaking, each activity equates to a chapter—except in the Publish phase, where we get a little more granular. These chapters are where we discuss the theories and detailed steps, showcase examples, and provide templates to equip you with the knowledge, skills, and tools to excel at your job.
- **Outputs**: These are the deliverables created as a result of executing each phase and activity. In the chapters of this book, we explain in detail how the outputs are a result of the practical steps and provide links to templates or samples you can use to create them.

7.2 [Theory] The Technical Writing Process

The Technical Writing Process

Phases	Plan	Design	Write	Edit	Review	Translate	Publish
Activities	☑ Collect Information, Data, and Knowledge ☑ Make a Plan ☑ Analyze Audience ☑ Define Review Team ☑ Estimate Scope, Time, and Cost* ☑ Develop Schedule*	☑ Design Structure (e.g., Table of Contents) ☑ Design Stylesheet ☑ Design Templates	☑ Write Drafts (First, Interim, Final) ☑ Include Images	☑ Edit Drafts	☑ Validate and Test Information ☑ Conduct Peer Review ☑ Conduct Subject Matter Expert Reviews	☑ Select Translation Partner* ☑ Translate and Localize Content* ☑ Create Terminology Database*	☑ Establish Document Control ☑ Obtain Approval ☑ Conduct Final Checks ☑ Publish Final Version ☑ Communicate with Stakeholders
Outputs	▪ Collected Information, Data, and Knowledge ▪ Documentation Plan ▪ Audience Profile / Personas ▪ Review Matrix in Documentation Plan ▪ Estimating Sheet* ▪ Project Schedule / Timeline*	▪ Document Structure (e.g., Table of Contents) ▪ Document Stylesheet ▪ Document Template	▪ First Draft Ready for Review ▪ Updated Document Structure / Table of Contents	▪ Edited Drafts	▪ Validated and Tested Information ▪ Request for Review ▪ Review Feedback ▪ Reviewed Draft	▪ Translated and Localized Content*	▪ Request for Approval ▪ Controlled and Published Document ▪ Message to Stakeholders

Manage

☑ Manage Progress	▪ Checklist ▪ Status Tracker ▪ Kanban Board ▪ Updated Project Schedule*

*Advanced or specialized topics / skills.

© 2024 Boffin Education™

Figure 14: The Technical Writing Process

Chapter 7 Tailor the Process

 What Does That Mean?

Deliverable

The output of an activity in a process or project. In technical writing, deliverables include user guides, manuals, and procedures—that is, content intended for use by end users—as well as internal documents, such as documentation plans and schedules, that are used in project planning and management.

Activity

A task, action, or milestone in a process or schedule. In technical writing, as in process analysis, activities are commonly expressed as verb-noun phrases—"define scope," "write first draft," and so on.

Process

A set of activities or tasks performed to accomplish an objective. Processes typically have a trigger that initiates the process; inputs necessary to perform the process; a corresponding result, or output; and a sequence of steps and decision points in between.

Note

Some Templates Are for Subscribers Only

Some of the templates in this book are for subscribers only. You'll need to subscribe to our online knowledge base at https://boffin.education/ to access them in editable format. Subscribers to the e-book and paperback versions of this book can use the discount code in Templates to obtain a free one-year subscription and access the full breadth of our technical writing content. If you don't want to subscribe, head over to our website. There you'll find many of the more straightforward templates available for free, and others are presented in a noneditable format as images.

7.2.1 Why Use the Process?

The Technical Writing Process is a highly practical, off-the-shelf methodology that can be customized and applied to any writing project, large or small. But it isn't just a toolkit—by marrying up the relevant theories that support each practical step, you can build your knowledge of your craft, as well as apply it in the real world.

Here are some benefits of using the process:

- **Time Efficiency**: The Technical Writing Process offers a ready-to-use framework that can potentially save hundreds of hours that would otherwise be spent in developing methodologies and tools from scratch.

- **Audience Focus**: This approach emphasizes the importance of understanding and addressing the specific needs of the audience, ensuring that the documentation meets their requirements.

- **Accessibility**: The book is designed to be easily understandable for newcomers and nonwriters. It explains technical concepts clearly and includes infographics for better comprehension.

- **Predictability**: By turning technical writing into a defined and manageable process, the Technical Writing Process reduces the reliance on ad hoc methods, increasing the predictability of project outcomes.
- **Consistency**: It provides a standardized method for content development that ensures uniformity across teams of technical writers, which aids managers in implementing consistent practices.
- **Planning Tool**: The process serves as a comprehensive guide for project managers. It helps to schedule and execute documentation projects by outlining high-level phases, outputs, and detailed activities.
- **Collective Wisdom**: The book is based on the experiences of its writers and includes insights and suggestions from a wide range of experts and seasoned writers, drawing from a wellspring of collective wisdom.

Introducing the process comes at an upfront cost—such as investing time in planning, and a more rigorous and structured way of working. However, the benefits of adding structure will quickly become apparent, as your projects produce better quality deliverables in a more consistent fashion.

 Insight

Measuring the Benefits

Technical writers often find that everyone wants the highest-quality content in the shortest amount of time. The art of technical writing is to do so in a way that creates the maximum value with the minimum of wasted effort and inconvenience to others. This is what the Technical Writing Process was designed for. Better yet, you can prove it with data! Consult our forthcoming book, *Technical Documentation Management*— available online at https://boffin.education/category/technical-writing/technical-writing-books/ for guidance on how to quantify the value and measure the success of your documentation projects.

7.2.2 What Documents Does the Process Apply To?

The Technical Writing Process applies to any type of document written by technical writers in print or digital format. These are discussed in detail in Chapter 2: Technical Writing Roles and Responsibilities | The Docs They Write. However, the principles in this book can be applied to virtually any project where the primary deliverable is documents (information).

 What Does That Mean?

Information

Data that has been processed and organized to support decision-making; for example, a document. Remember it as data "in formation."

Document

A discrete unit of information used to guide work, decisions, or judgment that serves as a guide to what should be done. Documents are forward-looking, as opposed to records, which are historical. Examples include technical documentation, plans, policies, and engineering drawings.

7.3 [Theory] Comparing Information-Development Frameworks

The Technical Writing Process is a high-level, sequential framework for developing information that has been specifically tailored for technical writing. JoAnn Hackos defines a similar process for developing information.[13] These different depictions aren't incompatible; on the contrary, they're well-aligned, albeit differently nuanced, perspectives.

Chapter 7 Tailor the Process

Where they differ is in the details, such as phase names, which activities are included in each phase, the names of the outputs for each phase, and so on. For example, we've split Manage and Translate into separate phases in our process. In Hackos's model, these are intertwined throughout the other phases, worked into the detail of the activities. The models also differ in focus Hackos includes an extensive focus on print production, while our process concentrates on digital production and leaves print production to the style guides and design manuals, where it's extensively covered. There are many other differences, which are too detailed and numerous to discuss here.

It's useful to have an understanding of both processes and how they align, as they're both popular, well-known methodologies in technical writing. In the accompanying diagram, we show how the high-level Technical Writing Process and the Hackos Information-Development Life Cycle overlap.

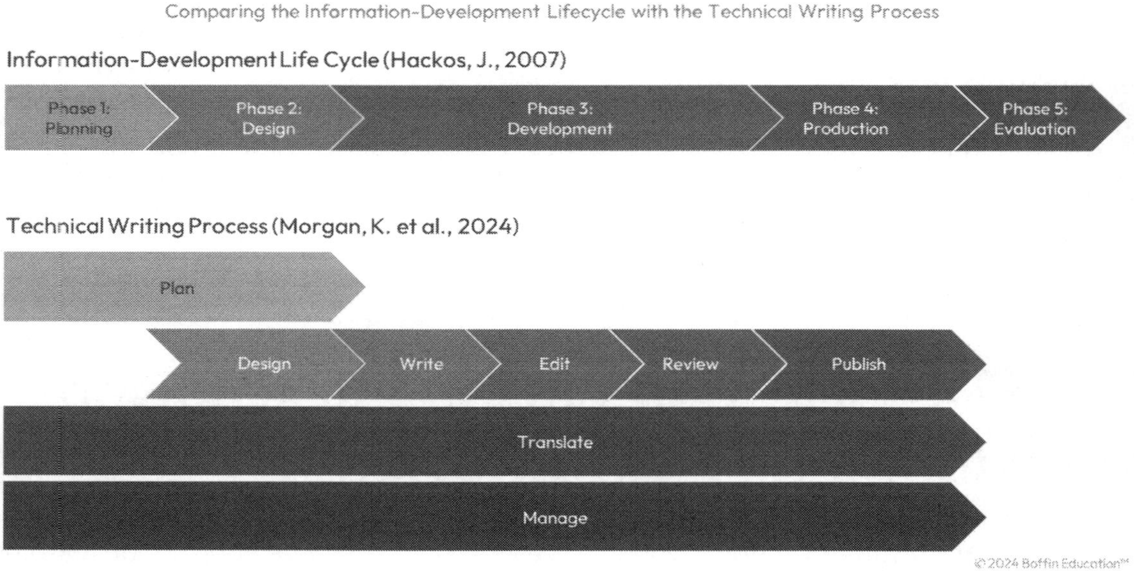

Comparing the Information-Development Lifecycle with the Technical Writing Process

Information-Development Life Cycle (Hackos, J., 2007)

| Phase 1: Planning | Phase 2: Design | Phase 3: Development | Phase 4: Production | Phase 5: Evaluation |

Technical Writing Process (Morgan, K. et al., 2024)

Plan

Design → Write → Edit → Review → Publish

Translate

Manage

© 2024 Boffin Education℠

Figure 15: Comparing the Information-Development Lifecycle with the Technical Writing Process

7.4 [Practice] Customize the Process

The Technical Writing Process is designed to be easily customized to suit your projects. In its comprehensive, end-to-end form, it's a robust methodology that supports even large, complex documentation projects. Before using it, we recommend modifying it to fit your project and your organization's unique culture, policies, and processes.

This section provides practical steps to show how this can be done successfully. We've also developed two case studies that showcase different implementations of the process.

 Tip

Some items in the Technical Writing Process—such as estimating and scheduling—are more advanced techniques, which aren't discussed in this book. If you're interested in learning more, check out our forthcoming title, *Project Management for Technical Writers*, on our website: https://boffin.education. Use the coupon code in Templates to claim your one-year free subscription to the site.

7.4.1 Step 1: Download and Learn the Process

Start by downloading the Technical Writing Process template. It's available on our website, https://boffin.education/, for free download: *Technical Writing Process Template*. Learn how it works by reading this book or taking one of our courses at https://boffin.education/technical-writing-courses/. Once you've built an understanding of the process, it's time for the analysis to begin.

7.4.2 Step 2: Analyze Your Organization's Requirements

Ensure you've developed a thorough understanding of your organization's unique requirements. Unfortunately, you may not always have the luxury of time to do so. Perhaps you've been brought in as a consultant or new manager with the expectation that you'll hit the ground running and quickly deliver results. Even so, take a few days to gather as much information as possible to build an informed understanding of the organization's needs. Don't just dive in! At the very least, consult some of the highly respected veterans—like senior writers—who have been with the organization for a long time and possess a wealth of accumulated wisdom.

For more guidance on information-gathering, see Chapter 9: Collect Information.

 Insight

Integrating Processes: Aligning with Organizational Dynamics

In addition to analyzing your organization's requirements, it's important to consider how your process may intersect with existing processes, such as Agile software development or product development processes. Speaking with product owners or project managers is important. These conversations can provide insights into their needs and expectations, clarifying how your process will integrate with the broader organization's dynamics.

7.4.3 Step 3: Add, Amend, Delete

Now it's time to customize the process. Begin with the macro view by considering the high-level phases in the Technical Writing Process. Some phases, such as Translate, might not be necessary for your project, so amend them as required. Then review the list of activities for each phase. Add, remove, or adjust as necessary. This stage is where you're likely to modify the process the most. Finally, review the list of outputs to ensure they support your now-modified process.

Use the Technical Writing Process Checklist to assist you in this journey.

 Tip

Balance Rigor with Flexibility in Your Process

You don't always need to define every micro step in the process. If you're leading a team of writers, leave some room for them to exercise their craft in a way that allows them to showcase their skills while achieving your desire for consistency. This is where the writers' adage "show, don't tell" comes into play. Show your fellow writers how to achieve excellence with time-saving templates rather than through lengthy rulebooks.

7.4.4 Step 4: Iterate and Improve

You probably won't nail the process on the first try. Processes are living frameworks—something that you will always be tinkering with and perfecting as the organization changes around you. If you're employed by your organization on an ongoing basis, rather than as a freelancer brought in for one-off projects, you can indulge your inner perfectionist by embracing the opportunity to polish the process. Schedule regular reviews—maybe every six to twelve months—to discuss the lessons learned from everyone's application of the process. Then feed their suggestions for improvement into the next iteration.

7.5 [Case Study] Case Study 1: Software Start-Up Company

 Voice of Practitioner

Francis

- **Role**: Technical writer with 4 years' experience
- **Location**: United Kingdom

What I Do: "I write for a fast-paced start-up that sells database software. We work in an Agile software development environment. My job is as a technical writer crafting online documentation for new features for end users."

Tools I Use: "For technical documentation, we use GitHub workflow, AsciiDoc, and Antora site generation software. We use the software development workflow process to produce documentation—that's kind of the higher-spec industry standard for good, future-looking companies."

A Day in the Life: "Morning time is spent checking my Slack and checking emails, and then I start solving the issue I'm on. Then I get on with the writing, getting feedback from subject matter experts, and updating my documentation. I spend a lot of time learning the product and understanding the product documentation, and learning the complicated parts of the business that others shy away from. I also spend about forty-five minutes a day at lunchtime learning how to code, which my company supports with a subscription to Codecademy."

What Does My Company Value? "When you need quick turnarounds in documentation, or if you're working with someone who doesn't have time for you, if you've spent your spare time understanding the product, you become a pretty safe person in the team and very valuable in the business."

7.5 [Case Study] Case Study 1: Software Start-Up Company

Francis's company, a start-up, is focused on quick releases and aligning documentation with its Agile development methodology. For this reason, Francis's planning consists of collecting information on new features, creating a list of topics aligned with sprints, and defining his review team. He's already familiar with his audience's needs and the purpose of his documents, having done this work already. Now he's relying on that knowledge to quickly move ahead with developing new topics. We've summarized Francis's process below. It's a lightweight version suitable for use in a fast-paced software development environment.

Table 8: Customized Technical Writing Process for a Software Start-Up Company

Phases	Customized Activities	Outputs
Plan	• Gather Software Feature Information	• Collected Information on New Features
	• Make a Plan	• List of Topics aligned with Sprints
	• Define Review Team	• Review Matrix for Topics
Design	• Design Structure	• Document Outline for Each Topic
Write	• Agile Sprint-Based Draft Writing	• Drafts of Topics
	• Incorporate Visuals	• Enhanced Topics with Screenshots
Edit	• Edit Topics	• Edited Topics
Review	• Validate and Test	• Validated and Accurate Topics
	• Conduct Peer Review	• Peer Review from Technical Documentation Team Members

Phases	Customized Activities	Outputs
	• Conduct Developer Review	• Developer Review Feedback
Publish	• Establish Document Control	• Controlled Document Versions on GitHub
	• Obtain Approval	• Approved Topics
	• Preview Documentation	• Site View on Antora
	• Synchronized Software Release Documentation Publishing	• Published Topics
	• Communicate with Stakeholders	• Slack Messages, Emails
Manage	• Manage Progress	• Daily Stand-Ups • Sprint Reviews • Jira Tickets Completed in Documentation Backlog

7.6 [Case Study] Case Study 2: Multinational Engineered Products Company

 Voice of Practitioner

Alison

- **Role**: Manager, technical communication support and processes with twenty years' technical writing experience
- **Location**: United Kingdom

What I Do: "I work for a multinational company and have a very varied role, which makes my job interesting. First, I work on customer-facing documents, so I create hardware documentation in both print and digital formats. The product line I'm responsible for includes engineered products such as servo drives and geared motors for machinery used in the food and beverage industry. I am also responsible for creating and improving all processes related to technical communication, such as the release of manuals and ordering illustrations and translations. I am the translation manager and also manage a team of four in India who support our department."

Tools I Use: "We use MadCap IXIA CCMS, both web and desktop clients. The web client has its own editor, and the desktop client uses Oxygen XML Editor. It's a long process—we're in the development of quite complex products here. The development from the initiation phase to market launch can take around two years."

Chapter 7 Tailor the Process

> **A Day in the Life**: "I generally start by looking through my emails and filtering out the easy ones. I'll look into my to-do list and set priorities for the day. If I've had a draft in review with engineering, product management, and hardware, I'll combine all the comments into a single PDF and go through it with them in a Zoom call. I'll check for what they want to do and validate any technical feedback, e.g., whether I have the correct values for something. Some things are easier to describe with graphics, so I might be reviewing that and placing a graphics order with our illustrator. I also check if any translation requests or illustration orders have been logged and distribute tasks to our support team in India. At the end of the day, I always add a comment 'up to here,' so I know where I'm up to."

> **What Does My Company Value?** "Learning how to juggle several tasks at once. You're not going to be working on one manual at a time. You might have one manual in review, be writing another one, and finishing a review task on another. Being able to juggle and organize tasks is key."

Alison's company is a large multinational engineered products company. Operating for almost a century now, it has multiple product lines across infrastructure, food supply, energy efficiency, and climate-friendly solutions. It still creates printed and shipped "in-box" documentation with its products—although it's moving away from this approach to one that focuses on creating content for multiple platforms, such as print, online, and chatbot. Alison's process focuses on creating technically accurate content supported by schematic diagrams, all published in multiple languages on multiple platforms. Even though her company is gradually moving away from printed manuals, they're still essential, so her process involves a heavy production component.

Table 9: Customized Technical Writing Process for a Multinational Engineered Products Company

Phases	Customized Activities	Outputs
Plan	• Collect Information, Data, and Knowledge	• Collected Information, Data, and Knowledge

7.6 [Case Study] Case Study 2: Multinational Engineered Products Company

Phases	Customized Activities	Outputs
	• Make a Plan	• Documentation Plan
	• Analyze Audience	• Audience Profile
	• Define Review Team	• Review Matrix in Documentation Plan
	• Estimate Scope, Time, and Cost	• Skeleton File (Table of Contents)
	• Develop Schedule	• Project Schedule
Design	• Design Structure	• Skeleton File (Table of Contents)
Write	• Write Drafts (First, Interim, Final)	• First Draft Ready for Review • Updated Table of Contents
	• Include Images	• Updated Document
Edit	• Edit Draft	• Edited Drafts
Review	• Validate and Test Information	• Validated and Tested Information
	• Conduct Peer Review	• Request for Review • Review Feedback
	• Conduct Subject Matter Expert Reviews	• Request for Review • Review Feedback

Chapter 7 Tailor the Process

Phases	Customized Activities	Outputs
Translate	• Approve Master Document/Content	• Approved Master Document
	• Define Target Languages	• Target Languages
	• Define Scope of Translation	• Translation Scope
	• Obtain Quotation	• Translation Provider Quotation
	• Translate Source Text	• Draft Translated Text
	• Edit and Proofread Translated Text	• Proofed Translated Text
	• Validate Translated Content	• Validated Translated Text
	• Layout Work (DTP—Desktop Publishing)	• Formatted Final Draft
	• Conduct Final Review	• Review Feedback
	• Deliver Final Format	• Final Draft Content on Translation Agency Site
	• Download Final Content and Import into Component Content Management System	• Final Draft Ready for Publication
Publish	• Establish Document Control	• Controlled Document
	• Obtain Approval	• Approved Document
	• Conduct Final Checks	• Final Draft Ready for Publication

7.6 [Case Study] Case Study 2: Multinational Engineered Products Company

Phases	Customized Activities	Outputs
	- Publish Final Version	- Published Document
	- Communicate with Stakeholders	- Email to Stakeholders
Manage	- Manage Progress	- Status Tracker - Updated Project Schedule

Part 3 Methods

Exploring the synergy between artificial intelligence (AI) and technical writing: enhancing efficiency while navigating new frontiers.

Chapter 8 AI for Technical Writers

Lead Writer: Caity Cronkhite | **Peer Reviewer/s**: Kieran Morgan | **Expert Reviewer/s**: Derek Moeller

This chapter examines the significant impact of artificial intelligence (AI) and large language models like ChatGPT on technical writing. It highlights AI's role in enhancing productivity by aiding in routine writing tasks and discusses its limitations in handling novel concepts. The chapter covers AI terminologies, its integration into writing processes, the importance of security and privacy in professional settings, and the role of technical writers in producing original, user-centric content.

 Who Should Read This

- Aspiring Technical Writers
- Beginner Technical Writers
- Career Advancers
- Managers of Technical Writers
- Cross-Domain Professionals
- Consultants

CONTENTS

8.1 Introduction	131
8.2 [Theory] What Is a Large Language Model (LLM) and How Does It Work?	132
8.3 [Opinion] Is Artificial Intelligence Going to Take My Job?	134
8.4 [Example] Using ChatGPT to Document a Payment Software API	136

Chapter 8 AI for Technical Writers

PROCESS

Inputs	Methods > Artificial Intelligence (AI) for Technical Writers		Outputs
<ul><li>Custom Instructions for ChatGPT</li><li>AI Prompts</li><li>Context for AI Prompts</li></ul>	**Theory**	• What Is a Large Language Model (LLM) and How Does It Work?	<ul><li>AI-Generated Content</li><li>AI-Generated Markdown</li></ul>
	Practice	• Worked Example: Using ChatGPT to Document a Payment Software API	
	Samples		
	Templates	• ChatGPT Prompt Library for Technical Writers	

8.1 Introduction

In late 2022, OpenAI released the first publicly available version of ChatGPT, an interface to a large language model (LLM) designed to generate humanlike text. Very quickly, this groundbreaking technology took the world by storm. In a matter of weeks, more than one million people downloaded ChatGPT, making it the "fastest-growing user application in history," according to research.[14]

Recent advances in artificial intelligence technology, particularly in large language models, have enormous implications for writers and communicators of all kinds. Humans are already leveraging artificial intelligence to write outlines, academic papers, blog posts, journalistic articles, and even full-length books. What impact, then, can we expect it to have on our roles and careers as technical writers?

In this chapter, we demystify some fundamental concepts of artificial intelligence and terminology. Through a worked example, we demonstrate how useful ChatGPT can be in streamlining typical writing tasks. For those wishing to explore more advanced uses, such as understanding the tool's limitations, finding solutions to typical issues, and applying ChatGPT to numerous other technical writing scenarios, we recommend our comprehensive resource: ChatGPT Prompt Library for Technical Writers.

Artificial intelligence isn't just a disruptive fad. When leveraged skillfully, it can be a powerful tool to boost productivity, eliminate the drudgery of some repetitive tasks, and increase your own efficiency as a technical writer.

8.2 [Theory] What Is a Large Language Model (LLM) and How Does It Work?

Artificial intelligence refers to the simulation of human intelligence by machines and computer systems. These computer systems are trained on massive amounts of data to simulate the process of learning, reasoning, problem-solving, perception, language, and decision-making.

Generative artificial intelligence is a subset of artificial intelligence models and the most recent major advancement in artificial intelligence technology. Generative artificial intelligence models are capable of generating new content—such as images, text, or music—that isn't directly copied from training data.

A large language model, or LLM, is a subset of artificial intelligence technology that's specifically designed to parse, understand, and generate humanlike text and content. Large language models are built on a deep-learning architecture and are trained on massive amounts of text data, including books, articles, websites, and, yes, even technical documentation.

You can interact with a large language model by entering a prompt into a platform or app, such as ChatGPT, that uses this technology. A large language model is a generative artificial intelligence model: it analyzes the context and patterns it learns from training data to predict the most appropriate next words or sentences, then generates a response that sounds like natural language.

8.2 [Theory] What Is a Large Language Model (LLM) and How Does It Work?

 What Does That Mean?

Artificial Intelligence (AI)

The simulation of human intelligence by machines and computer systems.

Generative Artificial Intelligence

A subset of artificial intelligence models capable of generating new content—such as images, text, or music—that isn't directly copied from training data.

Large Language Model (LLM)

A subset of generative artificial intelligence models designed to understand and generate humanlike text.

ChatGPT

A large language model, developed and released by OpenAI, designed to generate humanlike text and content in response to human prompts.

8.2.1 Dos and Don'ts for Using Artificial Intelligence at Work

Artificial intelligence tools have tremendous potential to help technical writers streamline their workflows and processes. However, tread carefully: the technology is still new, and many companies limit or prohibit the use of large language models for work purposes. Large language models, such as ChatGPT, store your input data for training purposes by default—although there are options not to do so, by purchasing premium or enterprise-grade plans. Here are a few dos and don'ts to keep in mind before you start using large language models in your own career.

Chapter 8 AI for Technical Writers

Table 10: Dos and Don'ts for Using Artificial Intelligence at Work

Do	Don't
Get permission from your employer before you use any large language model to write proprietary content.	Use public-facing large language models to write proprietary content unless you have express permission from your employer.
Review and fact-check content for accuracy and truthfulness—and to avoid plagiarism and bias—before publishing.	Include sensitive information, like trade secrets or customer data, in your prompts.
Make sure that using artificial intelligence in your work complies with legal regulations and data protection laws in your industry.	Publish content that was written by a large language model without human review for accuracy, truthfulness, and data compliance.

8.3 [Opinion] Is Artificial Intelligence Going to Take My Job?

Artificial intelligence technology, particularly large language models, is already dramatically changing the writing process and will continue to do so as the technology evolves. As we discuss in the next section, Chapter 8: Artificial Intelligence (A) for Technical Writers | [Example] Using ChatGPT to Document a Payment Software API, there are several ways that technical writers can leverage large language models to streamline common aspects of the writing process, from writing outlines to applying style-guide standards to our content. But can artificial intelligence replace technical writers completely?

Not so fast. Remember what we learned about large language models? Large language models are *generative* artificial intelligence models, which means they learn from existing content to predict what might come next in a sentence or paragraph.

8.3 [Opinion] Is Artificial Intelligence Going to Take My Job?

A large language model can easily produce information about products and concepts that have been written about extensively, assuming that information has been incorporated into the data used to train the model. However, a large language model can't formulate content or responses around ideas that have never been written about or incorporated into its training data set—at least not yet!

As technical writers, we're typically tasked with writing about new products, concepts, ideas, and technologies—a task that current large language models aren't programmed to do. In fact, the content that technical writers produce is likely to become even more important as artificial intelligence and large language model technology becomes more ubiquitous at work. The content we write about new products, services, and technology will likely become part of the corpus of information on which large language models are trained about new ideas. Large language models can't make decisions about what kind of content users might need, nor can they manage or maintain our content library for us.

Additionally, many technical writers create content for highly sensitive or proprietary products, data, or industries. Artificial intelligence and large language model technologies are relatively new in the public sphere, and there are widespread concerns about the security and safety of using these products to write content about highly sensitive information. Because most publicly available artificial intelligence products don't have established standards for protecting the data they receive in prompts[15], many companies and industries shouldn't use artificial intelligence or large language models to generate content at all at this time.

Artificial intelligence will certainly change how we do our jobs as technical writers, but it isn't going to replace us, at least not for now. By embracing this new technology and incorporating it into our work, we can streamline the technical writing process to make our work less tedious and produce better content for our users. Let's explore how.

8.4 [Example] Using ChatGPT to Document a Payment Software API

Let's take a look at a scenario where artificial intelligence might help us streamline various aspects of the technical writing process.

Imagine you're a technical writer working for an enterprise technology company that builds payment software. Your job is to document the company's APIs for external developers. You're the only writer assigned to the job, so you decide to use ChatGPT to help you streamline your writing process and save time.

> **Note**
>
> **A Note on Artificial Intelligence Preferences**
>
> ChatGPT isn't the only artificial intelligence large language model available on the market today. However, because of its ease of use and general accessibility, we refer primarily to ChatGPT in this chapter and elsewhere in this book.

8.4.1 Getting Started: Setting Up ChatGPT for Success

First you need to tweak a few settings to make sure that ChatGPT produces content that matches your audience's needs and that applies technical-writing industry standards and best practices. To increase our chances for success, we use ChatGPT's custom instructions feature.

Custom instructions allow us to:

- Provide information about who we are, our role, and the kind of content we're working on.
- Define the audiences we're writing for.

8.4 [Example] Using ChatGPT to Document a Payment Software API

- Include specific guidelines and instructions for the content ChatGPT produces, such as voice and tone, punctuation, or style guidelines.

Here's an example of the custom instructions you might use for your technical writing role:

Chapter 8 AI for Technical Writers

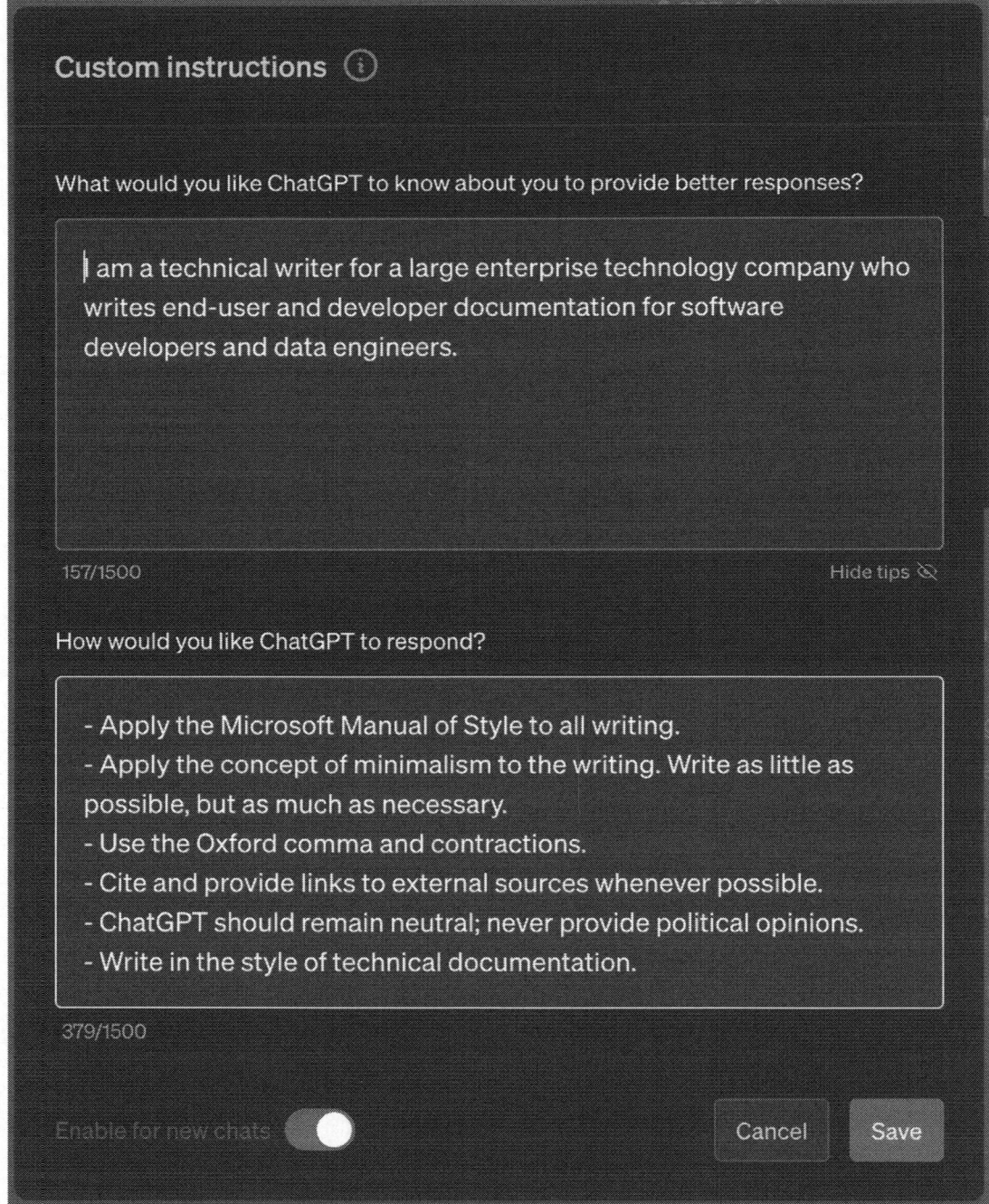

Figure 16: Example of Custom Instructions in ChatGPT

8.4.2 Use Case 1: Content Planning

Now that you've set your custom instructions, it's time to get to work planning the documentation for a new API your company is working on. You want to outline the structure and information architecture of the documentation before you start writing.

Chapter 8 AI for Technical Writers

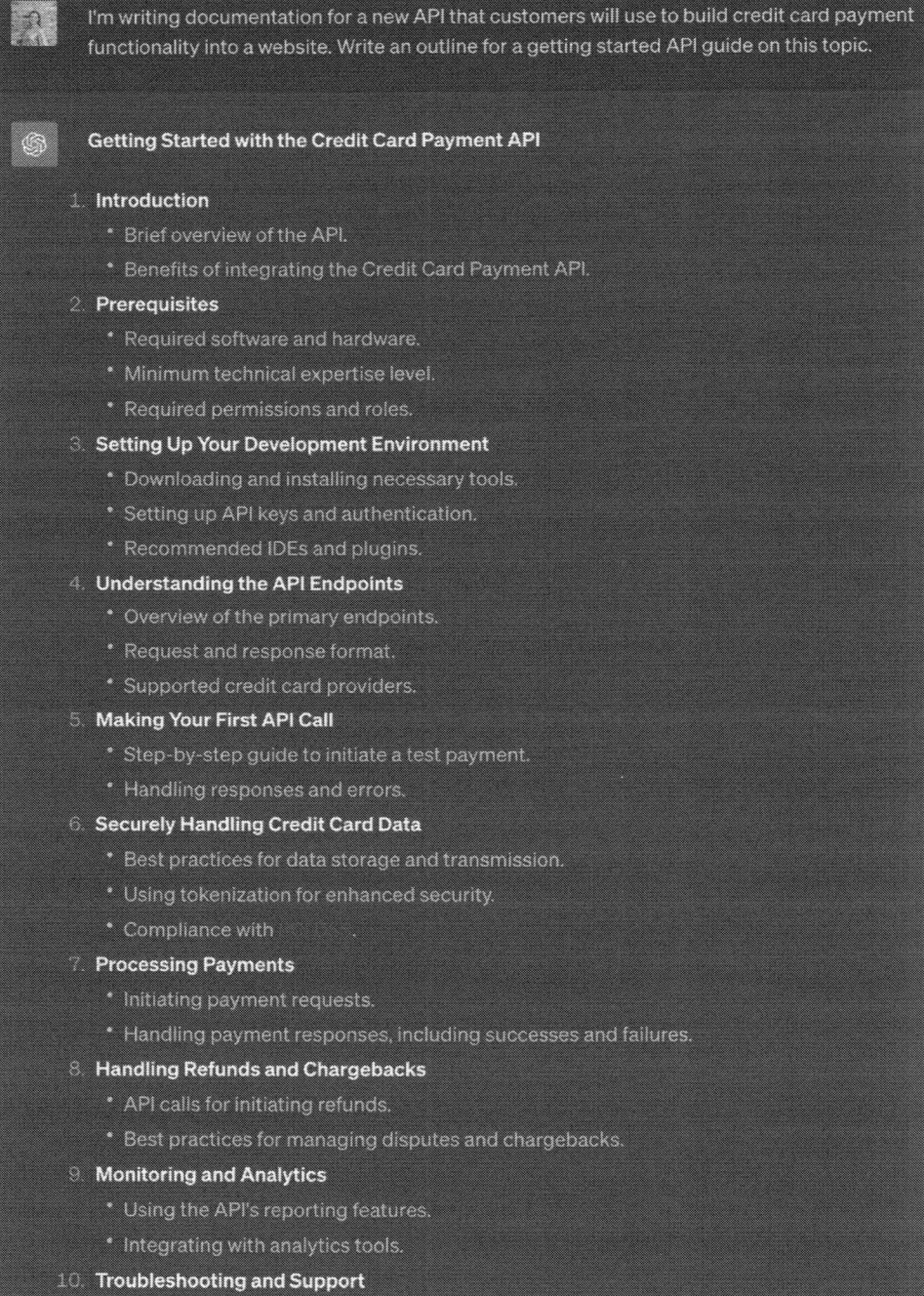

Figure 17: Example of Content Planning in ChatGPT for Technical Writing

8.4 [Example] Using ChatGPT to Document a Payment Software API

With a simple prompt (and the custom instructions you provided earlier), ChatGPT produces an outline of basic topics you'd likely want to include in API documentation for a feature like this.

You also want to provide some useful examples and developer tutorials. ChatGPT can help with that too!

Chapter 8 AI for Technical Writers

 That's a great start. Help me brainstorm ideas for sample applications that a developer might want to build with this API.

 Certainly! Here are some sample applications developers might consider building with the Credit Card Payment API:

1. **E-commerce Platform**
   - Enable online merchants to list products and process credit card payments securely.
2. **Subscription Management System**
   - Recurring billing for services like software, digital content, or memberships.
3. **Point of Sale (POS) System**
   - A web-based checkout solution for physical stores or pop-up shops.
4. **Invoice and Billing Solution**
   - Generate and send invoices to clients, with an option to pay directly via a link.
5. **Ticketing System**
   - For events, theaters, or transportation, allowing customers to purchase and receive tickets.
6. **Donation Platform**
   - Facilitate charitable donations with options for one-time or recurring contributions.
7. **Booking and Reservation System**
   - For hotels, restaurants, or service appointments, allowing secure payment upon reservation.
8. **Mobile Wallet Integration**
   - Add credit card payment functionality to a mobile wallet or payment app.
9. **Digital Marketplace**
   - A platform where digital goods (e.g., e-books, software, digital art) can be sold.

Figure 18: Example of Brainstorming with ChatGPT

8.4.3 Use Case 2: Writing Content

Planning content is great, but you have deadlines to meet, and you need to start writing.

Remember, ChatGPT only has access to information and data that's publicly available on the web. Your company's API hasn't been released yet, so ChatGPT can't write factual information about your API and its features. However, if prompted, it will certainly generate plausible-sounding—and possibly wildly inaccurate—content.

Luckily there are a few topics in your documentation outline that are more generic, such as Section 6: Securely Handling Credit Card Data. You use ChatGPT to write a first draft of a topic that describes what PCI DSS is and why it matters. Here's a snippet:

Introduction

The Payment Card Industry Data Security Standard (PCI DSS) is a set of security standards designed to ensure that all companies that accept, process, store, or transmit credit card information maintain a secure environment. Understanding and adhering to these standards is crucial for any business dealing with cardholder data.

What is PCI DSS?

PCI DSS stands for Payment Card Industry Data Security Standard. It's a global framework that provides businesses with guidelines to handle cardholder information in a secure manner, reducing the risk of financial data breaches.

Key Components of PCI DSS

PCI DSS comprises 12 primary requirements, categorized into six control objectives:

- **Build and Maintain a Secure Network**
    - Install and maintain a firewall.
    - Change vendor-supplied defaults for system passwords and other security parameters.
- **Protect Cardholder Data**
    - Protect stored cardholder data.
    - Encrypt transmission of cardholder data across open networks.
- **Maintain a Vulnerability Management Program**
    - Use and regularly update anti-virus software.
    - Develop and maintain secure systems and applications.
- **Implement Strong Access Control Measures**
    - Restrict access to cardholder data to need-to-know only.
    - Assign a unique ID to each person with computer access.
    - Restrict physical access to cardholder data.
- **Regularly Monitor and Test Networks**
    - Track and monitor all access to network resources and cardholder data.
    - Regularly test security systems and processes.
- **Maintain an Information Security Policy**
    - Establish, publish, maintain, and disseminate a security policy.

Why PCI DSS Matters

Figure 19: Using ChatGPT to Write a First Draft of a Technical Document

Not a bad start!

(Keep in mind that, although that first draft looks pretty good, you always need to verify the information to make sure it's accurate and factual before you publish to your audience.)

 Tip

Leverage ChatGPT's Technical Writing Abilities

Many folks aren't aware that GPT-4 can translate code into human language explanations—with some limitations. This capability can be extremely useful for technical writers working on developer documentation. For instance, an API's code endpoints, complete with type hints and function docstrings, can be fed into GPT-4, and it will produce a rough draft explanation of the API, complete with prompts. Because the accuracy of ChatGPT's explanations varies, the rough draft must be reviewed and validated by subject matter experts—a process that you, as the writer, are responsible for orchestrating. Before doing so, remember you should never put your organization's code into ChatGPT without first getting permission from your employer.

8.4.4 Use Case 3: Content Editing, Formatting, and Conversion

Your company publishes documentation in Markdown, so you need to convert the content ChatGPT wrote for you into the Markdown format. Never fear: ChatGPT is adept at that too.

Chapter 8 AI for Technical Writers

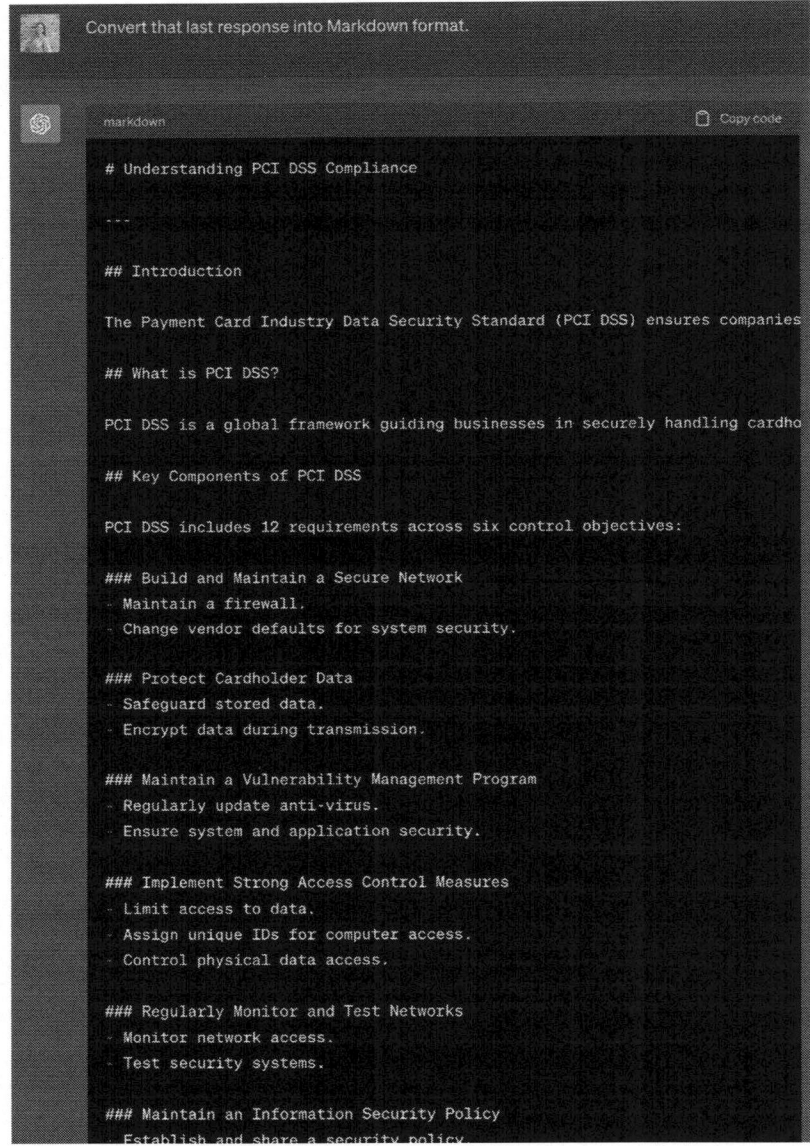

Figure 20: Using ChatGPT to Convert Content into Markdown Format

In less than a minute, ChatGPT tags the content accurately, allowing you to copy and paste it seamlessly into your existing Markdown content.

Part 4 Plan

Strategies and frameworks for planning successful technical documentation projects.

Chapter 9 Collect Information

Lead Writer: Kieran Morgan | **Peer Reviewer/s**: Felicity Brand | **Expert Reviewer/s**: Dina Bennett

This chapter equips you with essential steps to gather the information you'll need before diving into the creation of technical documents. We introduce the DIKW Pyramid as a conceptual framework to understand the relationships between data, information, knowledge, and wisdom. Furthermore, the chapter provides insights on locating key resources, such as brand guidelines, templates, and style guides. Finally, we offer tips to set the stage for a successful start to any new documentation project.

 Who Should Read This

- Aspiring Technical Writers
- Beginner Technical Writers
- Career Advancers
- Cross-Domain Professionals

CONTENTS

9.1 Introduction 151
9.2 [Theory] DIKW Pyramid 151
9.3 [Practice] Collect Information, Data, and Knowledge 154

Chapter 9 Collect Information

PROCESS

Inputs	Plan > Collect Information		Outputs
• Technical Product Data • Information from Existing Documents • Knowledge from Subject Matter Experts (SMEs) • Access to Hardware and Systems • Style Guides	**Theory**	• DIKW Pyramid	• Collected Information, Data, and Knowledge
	Practice	• Check for Templates or Style Guides • Secure Access to Hardware and Systems • Check Out Competitors' Documentation • Check for Standards	
	Examples	• N/A	
	Templates	• Resource Checklist for Technical Document Creation	

9.1 Introduction

Technical writers are like proverbial crows: we gather and hoard information with such dedication that you'd be forgiven for thinking we're going to decorate our nests with it. Don't be alarmed! This is completely normal behavior for technical writers and something we should all be doing—not only at the start of a project, but at every point along the way.

Technical writing is all about synthesizing a focused piece of information—a technical document—from a variety of sources. This process creates knowledge in the minds of users and enables them to carry out tasks they may have previously struggled with.

9.2 [Theory] DIKW Pyramid

Although most people use the terms "data," "information," "knowledge," and "wisdom" interchangeably, the data, information, knowledge, and wisdom (DIKW) Pyramid provides us with more precise language to articulate our craft.

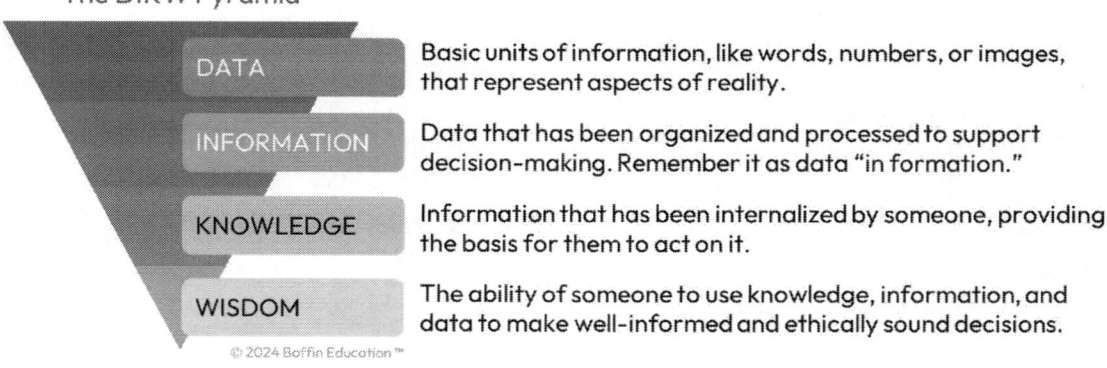

Figure 21: The DIKW Pyramid

Let's imagine you're a technical writer tasked with writing a user guide for a power saw. How would you explain that in terms of the DIKW Pyramid?

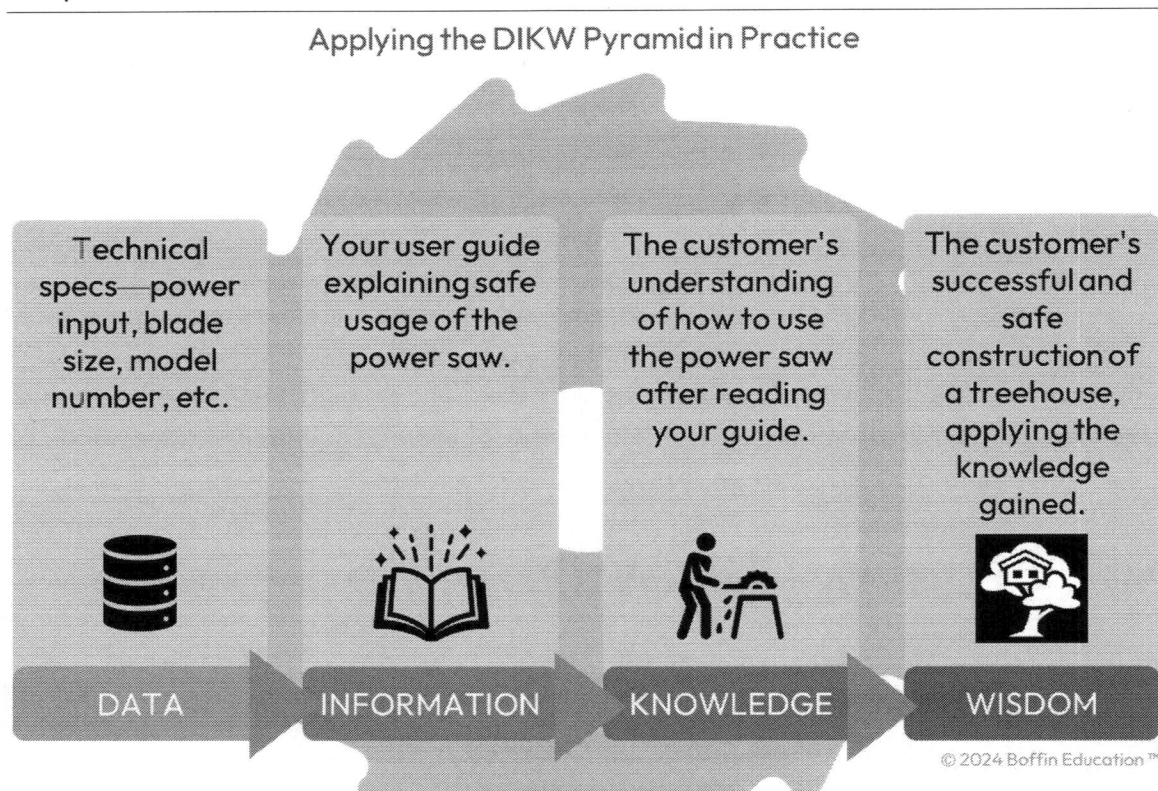

Figure 22: Applying the DIKW Pyramid in Practice

By understanding the DIKW Pyramid, you gain a structured language for discussing and refining your craft, ensuring that you not only collect and hoard information like our metaphorical crow, but also use it to create documents that transform data into actionable wisdom.

9.2 [Theory] DIKW Pyramid

 Note

Is This the DIKW Pyramid You Remember?

There are many variations on the DIKW Pyramid, and there's no rule on which one to use. In fact, some of them even add additional layers, such as "content."[16] We've synthesized our own DIKW Pyramid, drawing from Andrew Liew and Jennifer Rowley's excellent definitions.[17, 18]

 What Does That Mean?

Data

Basic units of information, like words, numbers, or images, that represent aspects of reality.

Information

Data that has been processed and organized to support decision-making; for example, a document. Remember it as data "in formation."

Knowledge

Information that has been internalized by someone, providing the basis for them to act on it.

Wisdom

The ability of someone to use knowledge, information, and data to make well-informed and ethically sound decisions.

9.3 [Practice] Collect Information, Data, and Knowledge

It's common for technical writers, especially freelancers or contractors, to begin a new project with limited clear instructions regarding expected deliverables. Your manager might not even be fully aware of what the required deliverables should look like; they only know documentation will be needed. This situation is normal, so your first task on a new project should be to activate your inner "bird mode." Start by searching for existing information before diving into writing or planning. There's often existing content that you can leverage or repurpose to create your own documentation.

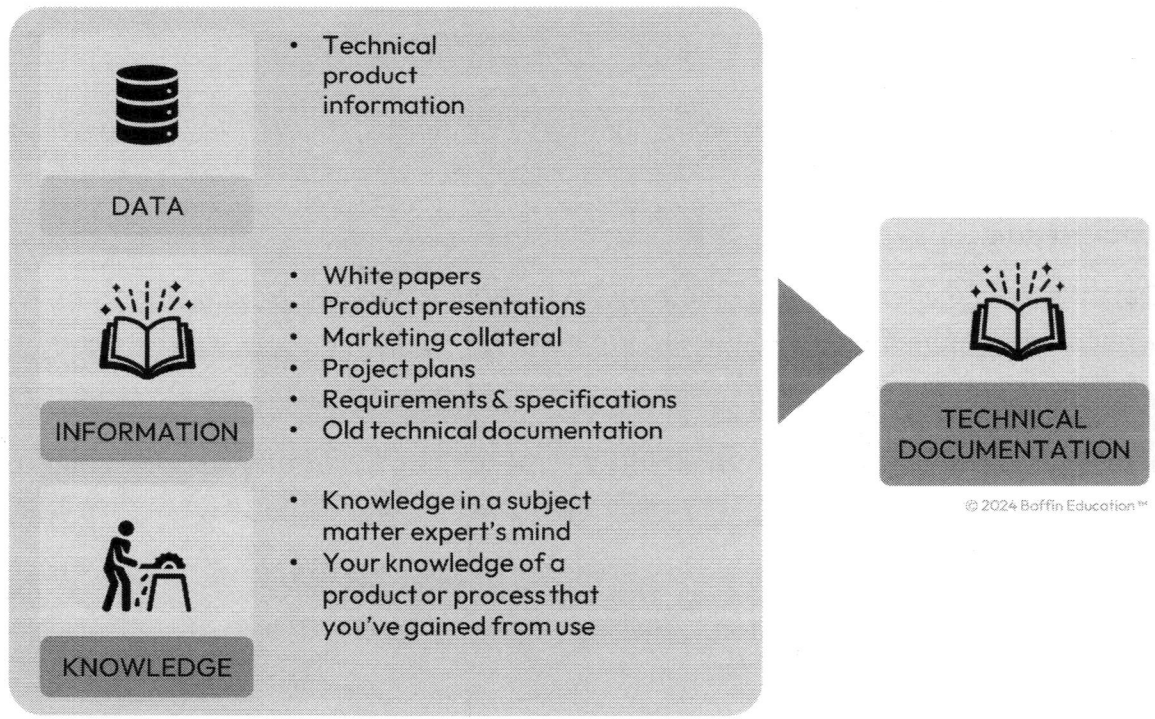

Figure 23: How Data, Information, and Knowledge Become Technical Documentation

If you're documenting a product (hardware or software):

- Don't limit your research to purely technical content or experts. Consult both product managers and engineers. They often have high-level documents, such as board presentations or requirements documents for development. These documents will outline nontechnical aspects, such as the problem the product aims to solve, profiles of intended users, wireframes, and visual mock-ups.

If you're writing procedural documentation:

- Consult the subject matter experts who perform the tasks you're documenting. These individuals often have their own documentation which was created as memory aids or for other purposes. Even if they don't have documentation, build relationships with them early on, as you'll need their expertise later. Be cautious when reviewing any documents they provide; they might be outdated or of questionable quality, which is likely why a technical writer is needed.

 Insight

What If There Is No Existing Information?

Occasionally, you might find that no relevant information exists. While this is common for process and procedure projects, it's a red flag for product development projects. How can a product be planned and developed without specifications? If this happens, discuss the situation with your manager and diplomatically explore why existing documentation is lacking. You might also inquire discreetly within the organization to uncover any hidden information. If you discover that no information exists, consider it an opportunity. Teams responsible for creating this information, such as marketers and engineers, may need your technical writing expertise, allowing you to contribute your skills to the project.

9.3.1 Check for Templates or Style Guides

It's crucial to check for existing templates, style guides, and brand guidelines before you write. In larger organizations, templates will often be available that you can—or must—use or adapt for your technical documents. There will also often be organization-specific style guides or brand guidelines that provide guidance on how to write in the company style, properly acknowledge company trademarks, and so on.

These in-house style guides differ from general style guides like the *Chicago Manual of Style*. General style guides contain broadly accepted rules of writing and grammar. They're designed to be all-purpose or applicable to a particular industry or region. Organization-specific style guides either supplement or overrule elements of these general guides to align the company's brand voice as consistently as possible when communicating with customers.

 Insight

Insight: Turn a Lack of Guidance into an Opportunity

If your organization doesn't have style guides or templates for the documents you'll be writing, don't despair. This situation can be an opportunity for you to add value by creating them yourself—provided your manager approves. Style guides and templates save time on future projects, ensure consistency, and create a professional look that aligns with the organization's brand. They're also an excellent opportunity to enhance your skills, making you more employable in the future. Creating templates and style guides is a senior technical writing skill.

9.3.2 Secure Access to Hardware and Systems

One task you'll need to sort out early on is access to the item you're documenting. Whether it's hardware or software doesn't matter; you'll create more accurate documentation if you can test it against your instructions. Of course, that's not always practical or feasible, but you should aim for it.

Whether you're documenting hardware or software, follow one golden rule: submit your request as early as possible. Gaining access to working prototypes or software environments where you can experiment without breaking anything can take time.

If you're documenting processes:

- Processes often involve some element of human interaction with software, hardware, or both. You may not always need access to these systems, but you will need access to subject matter experts to validate your documents.

- If they've agreed to demonstrate a procedure, request in advance that you can take frequent screenshots from their screens as they proceed. Ask for their patience if you need to slow them down to capture everything thoroughly.

If you're documenting hardware:

- If possible, get involved early in the product development lifecycle to be integrated into any usability testing conducted by the design team.

- Ask the engineering team for access to a working prototype to test against your documentation. This is crucial for ensuring your documentation is accurate.

- If your hardware interfaces with software, also secure access to that software, including if possible any testing and diagnostic software used to troubleshoot the device.

Tip

What If There's No Working Prototype?

Sometimes there may not be prototypes available, or they might not be accessible within the limited time you have to document them. In such cases, work closely with engineers to understand your alternatives, such as virtual simulations or CAD drawings.

If you're documenting software:

- Request software access as early as possible. You'll need it to understand how the software works, capture screenshots, and validate workflows. You might also discover bugs and usability issues.

- Consider the permission levels you'll need. Administrator access will allow you to access all features and create or delete any dummy data needed for testing and documentation. However, the standard user view might differ significantly, so make sure you cover all your bases.

- Ensure the version you access is regularly updated to reflect the user experience accurately. Maintain regular conversations with developers to keep them aware of your activities—just in case they forget to update the environment in which you're working.

 Tip

Make Sure You're Documenting the Correct Version

When documenting a system, ensure you're working with the correct version. Features and workflows can change dramatically with updates, quickly rendering your documentation obsolete. Maintain regular, proactive conversations with developers to understand if they've updated the code base for your documentation environment and when features will be stable enough for documentation.

Most development teams will have several environments they work in:

- **Production or "Live" Environment**: This is the environment that end users and customers interact with. As a technical writer, you most likely won't need access to the production environment.

- **Staging or "Sandbox" Environment**: This is a close replica of the production environment but separate and isolated from it. Developers use this environment to test code changes.

- **Test or User Acceptance Test (UAT) Environment**: Like the staging environment, this is a close replica of the production environment but specifically set aside for testing purposes to identify bugs or usability issues in new or updated features.

- **Local Environment**: You may have the opportunity to install and run the software on your own computer, allowing you to test the software as if you were an end user.

 Note

What to Do If Access Is Denied?

It's not uncommon for technical writers to face resistance when requesting access. Various reasons, such as security concerns, can account for this. However, it's crucial to be both polite and persistent. Here are some steps to consider:

- Ensure you've used any formal access request procedures so your request is officially on the record.

- Request an appropriate access level that minimizes risks, such as exposing sensitive customer information.

- Identify a subject matter expert who has access and ask if you can take screenshots or photographs.

- If denied access, consider escalating your concerns to your manager, project manager, or even the legal team.

9.3.3 Check Out Competitors' Documentation

If competitors publicly share their documentation, don't hesitate to review it. This can provide valuable inspiration and insights into industry norms for various types of technical content.

> **Note**
>
> **Acknowledging Other People's Work**
>
> Always be aware of copyright laws. If you intend to repurpose someone else's content, you should acknowledge it appropriately. For specific guidance, refer to your style guide, which should include details on copyright and acknowledgment practices.

9.3.4 Check for Standards

In technical writing, adhering to specific standards is sometimes a necessity. Standards, including those established by bodies such as the International Organization for Standardization (ISO), the Institute of Electrical and Electronics Engineers (IEEE), and Aerospace and Defense industries (ASD), provide a framework for content and presentation that aligns with industry norms, best practices, and regulatory requirements. Before taking the plunge and starting to write, it's best to check whether your documentation is expected to align with a particular standard.

Why Standards Matter

1. **Compliance and Quality Assurance**: Standards like ISO 9001 for quality management and ASD-STE100 for simplified technical English help to maintain a high level of quality and consistency in technical documentation. Compliance with these standards ensures that the documents meet specific benchmarks set within the industry.

9.3 [Practice] Collect Information, Data, and Knowledge

2. **Legal and Regulatory Requirements**: In certain industries, following specific standards is legally mandated. This is particularly relevant in fields with high safety and regulatory requirements, such as aerospace, healthcare, and engineering.

3. **Best Practices and Benchmarks**: Standards often embody best practices within an industry. For example, ISO/IEC/IEEE 26515:2011 provides guidelines for developing user documentation in an Agile environment, which can be a valuable resource for teams working in Agile methodologies.

Translation and Localization Standards: ISO 17100:2015 sets comprehensive guidelines for translation services, ensuring quality and consistency in multilingual technical documentation. This standard is pivotal when preparing documents for diverse linguistic audiences. See *Part 9: Translate*.

 What Does That Mean?

ISO 9001

ISO 9001 is the global standard for quality management systems. It's published by the International Organization for Standardization, a nongovernmental organization that develops and publishes international standards across a wide range of industries and sectors. This standard focuses on ensuring companies meet customer needs for their products and services. It also sets out requirements for document control, which we discuss in depth in Chapter 24: Publish.

Chapter 10 Make a Plan

Lead Writer: Kieran Morgan | **Peer Reviewer/s**: Steve Moss

This chapter guides you through the process of planning documentation projects using the PADRE method (Purpose, Audience, Design, Review) for effective project execution. The chapter underscores the significance of planning in ensuring project success, particularly in Agile environments, and provides practical tools like checklists and templates. It emphasizes the need for proper planning to create a strong foundation for the success of your technical documentation projects.

 Who Should Read This

- Aspiring Technical Writers
- Beginner Technical Writers
- Career Advancers
- Managers of Technical Writers
- Cross-Domain Professionals
- Consultants

CONTENTS

10.1 Introduction	165
10.2 [Theory] The PADRE Method	167
10.3 [Practice] Make a Plan	171

Chapter 10 Make a Plan

PROCESS

Inputs	Plan > Make a Plan		Outputs
• Collected Information, Data, and Knowledge • Initial Briefing • Stakeholder Meetings	**Theory**	• The PADRE Method: Purpose, Audience, Design, Review	• Documentation Plan
	Practice	• Choose the Right Plan • Leverage Initial Managerial Briefing • Develop Plan • Refine Plan • Approve Plan	
	Examples	• Example of the PADRE method in practice	
	Templates	• Briefing Checklist for Technical Writers • Documentation Plan Template • Documentation Microplan Template • Documentation Microplan Template for Agile Projects	

10.1 Introduction

Ever felt like you're embarking on a technical writing project with just a vague notion of what's expected? You're not alone. Many technical writers begin with minimal guidance and often dive into projects where even managers or organizations haven't fully grasped the scope of the needed deliverables.

This is where it pays to make a plan. Plans aren't just documents; they're like contracts of expectations between you and your organization or client that outline precisely what's to be delivered and when. Beyond that, they're your tool for rallying support, gaining buy-in, and guiding stakeholders through the project's journey.

In this chapter, we'll guide you through the process of how to make a plan. To keep things manageable, we've provided options: a streamlined microplan and a more traditional, comprehensive documentation plan. We've also provided a tailored option for Agile projects.

Both approaches use our variation of the PAD (purpose, audience, design) method, which is a relatively new method that's quickly becoming widespread. We call this the PADRE method: purpose, audience, design, and review. We've created templates for you to easily apply the PADRE method to your own documentation projects.

By following the instructions in this chapter, you'll be able to select a planning approach that works best for both your project and your skill level. This will get your documentation project off to a flying start in the minimum possible time.

10.1.1 Do I Really Need a Plan?

Many technical writers feel the urge to show productivity straight away by diving straight into documenting. Does it really pay to plan? In fact, doesn't the Agile Manifesto state, "Responding to change over following a plan"?[19] In short: Yes, it pays to plan, even in an Agile project. There are two reasons:

Chapter 10 Make a Plan

1. **Planning creates consensus**: Creating a plan is a great way to achieve consensus among your stakeholders on the assumptions you've made about your project. This reduces the likelihood of surprises later down the track—for example, if it turns out an assumption has been made that's later proven to be incorrect. In this way, the process of arriving at a plan irons out many of the potential problems or misunderstandings that will inevitably arise later in the project.

2. **Planning is linked to success**: Studies on successful projects clearly show that more time spent planning leads to better results.[20] Having said that, as with any endeavor, it's important to balance preparation and execution without going overboard and over-engineering things. That's why we've provided options that range from essentials through to advanced for you to use when planning your projects.

As a writer, the pressure to get down to writing straight away can come from your own innate desire to quickly prove your value, or it can be imposed by others who only see the value in your final output rather than in the planning and preparation you put in behind the scenes. Remember that it pays to do a thorough—but not excessive—job in planning and preparation, and that it's okay to diplomatically but firmly communicate that need to others.

 Note

Planning for Agile Projects

Agile projects generally have shorter development cycles, and priorities can quickly change as the product team iterates on features. Does it make sense to have a plan in this case? Absolutely yes! The very nature of Agile projects makes it difficult to keep track of what is being developed and when. Having a plan that shows how specific types of documents align with user stories and development sprints goes a long way in establishing trust in your technical writing process. In the case of Agile projects, your plan does not need to be extensive, so we've created a customized version of the plan template for you to use.

10.2 [Theory] The PADRE Method

The PAD method is a relatively new technique for planning technical documentation projects that's rapidly gaining popularity.[21] Why? Because it's simple—and memorable! It boils down the essence of a technical writing project into three key dimensions: purpose, audience, and design of deliverables.

We've created our own version of the PAD method, complete with templates so that you can easily apply it. We call it the PADRE method because it incorporates an additional, but essential, element of planning: the review team. Defining a review team is critical. It's only through a thorough review process that you can ensure your documentation is high quality and fit for use by its audience.

Here are the key elements of the PADRE method:

- **Purpose**: What the document aims to do.
- **Audience**: Who will read it and why they need it.
- **Design**: What the final documents will look like and when they will be ready.
- **Review**: Who will review and approve your documents to ensure quality and accuracy.

It doesn't matter whether your project is straightforward or complex—the PADRE method still applies. It's the core element in all our documentation plan templates.

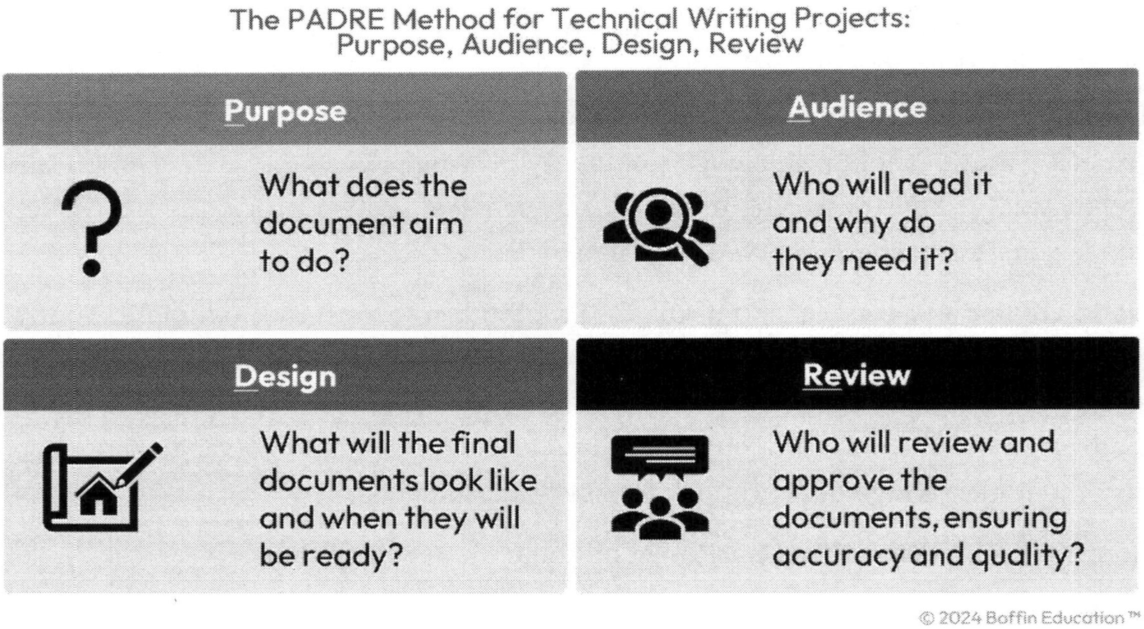

Chapter 10 Make a Plan

The PADRE Method for Technical Writing Projects:
Purpose, Audience, Design, Review

Purpose	**Audience**
What does the document aim to do?	Who will read it and why do they need it?
Design	**Review**
What will the final documents look like and when they will be ready?	Who will review and approve the documents, ensuring accuracy and quality?

© 2024 Boffin Education™

Figure 24: The PADRE Method for Technical Writing Projects: Purpose, Audience, Design, Review

Below is an example of the PADRE method in action, based on a fictional scenario: a technical writer developing a suite of documentation for an intraoperative medical device.

Table 11: Example of the PADRE Method in Practice for Medical Device Documentation

Purpose	The primary purpose of the documentation is to provide accurate and easily understandable instructions for surgeons to operate the intraoperative device and its connected software safely and effectively.

Audience	**Primary Audience**: • The primary audience comprises surgeons who will be using the intraoperative device and its connected software during procedures in an operating theatre. **Secondary Audience**: • The secondary audience includes nurses who assist the surgeon by preparing and holding the equipment during the procedure.
Design	The final output should include specific types of documentation tailored for both the primary and secondary audiences, each with its respective format and content. **Primary Audience Deliverables**: • A comprehensive user manual in PDF format for surgeons that details all functionalities, safety measures, and step-by-step procedures. • Interactive video tutorials that are accessible through a secure portal. **Secondary Audience Deliverables**: • A quick-start guide in laminated paper format to be kept in the operating theater, to aid nurses in preparing and holding the equipment. • Checklists for setup and maintenance to be included in the nurse training modules.

Review	This section details the teams responsible for reviewing the document for quality and accuracy. **Reviewers**: - **Surgical Team**: Surgical procedures and medical terminology. - **Nursing Team**: Equipment setup and nursing procedures. - **Tech Team**: Software interface and connectivity. - **Tech Docs Team**: Adherence to best-practice document design. **Approvers**: - **Quality Team**: Sarah Brown, Quality Manager - **Medical Board**: Dr. Alan Watts, Board Chairman - **Legal**: Laura Williams, Legal Advisor

 **Note**

The Ancient Roots of PAD

The PAD method is a rhetorical analysis technique. Rhetoric is the art of effective communication and persuasion. It involves connecting with your audience in a way that encourages them to see things from your perspective. This is an essential skill for any technical writer! By connecting and empathizing with your audience, you enhance your ability to convey complex ideas. Although quite different from its original form, the PAD method has its origins in Aristotle's teachings.[22] Aristotle was a Greek philosopher who lived from 384–322 BCE and made significant contributions to numerous fields, including developing the concept of rhetoric.

10.3 [Practice] Make a Plan

The following steps take you through the process of making a plan that's right for your project. Whether it's straightforward, Agile, or a large, complex waterfall-style project, we've got you covered.

 Note

Avoid a "Tick-the-Box" Mentality

Try to avoid a tick-the-box mentality while you're making your plan. Remember, the aim of a plan isn't simply to populate a template. Creating a plan is how you arrive at a comprehensive understanding of what you're supposed to be delivering, why you're delivering it, and who you're delivering it to.

10.3.1 Step 1. Choose the Right Plan for Your Project

The first step in making a plan is to select the plan that's suitable for your project's scope and complexity. To make your choice easier, we've developed customizable templates for several common scenarios.

Table 12: Different Documentation Plan Options for Different Projects

Plan	Description	When to Use
Documentation Microplan Template	A streamlined alternative to traditional documentation plans designed to quickly guide technical writers through the planning stages of a new documentation project. It uses the PADRE method to analyze purpose, audience, design of deliverables, and the review team responsible for ensuring quality and accuracy.	• When you anticipate a relatively straightforward project. • When you've been working at an organization for a while and all the usual stuff you'd put in a plan is by now ingrained knowledge. • When you just don't have the time available to make a detailed plan.
Documentation Microplan Template for Agile Projects	A customized version of the microplan specifically tailored for Agile projects. It's designed to align with Agile workflows and sprints, helping technical writers quickly navigate through the planning stages of a new documentation project.	• When you're working on an Agile software development project.

Plan	Description	When to Use
Documentation Plan Template	A comprehensive planning tool that considers multiple variables in addition to PADRE to create a holistic, integrated plan. Similar to a Project Management Plan used by project managers but tailored for technical documentation projects.	• When you want to thoroughly plan for all contingencies in advance, including such things as review and approval responsibilities, lead times, dependencies, assumptions, and constraints. • When you have a more complex project with multiple deliverables. • When you're unfamiliar with the context or organization and you need to thoroughly understand things before proceeding. • When you're a consultant developing a scope of work for a client and you need everything thoroughly documented so it can form part of a contract.

10.3.2 Step 2. Leverage Your Initial Briefing with Your Manager

Your journey as a writer on a new project—or starting a new role—often begins with a one-on-one meeting between you and your manager. This initial briefing is generally set up by your manager, but if not, take the initiative and propose a meeting yourself. During this session, your manager will typically outline their vision for the documentation and provide a wealth of information about the project. You can expect to receive details about the project's purpose, the key stakeholders involved, the expected timeline, and where to locate essential documents.

Use this meeting as an opportunity to ask questions. Your manager will expect you to be proactive in seeking clarification on any ambiguous points. Use the Briefing Checklist for Technical Writers as a prompt to make sure you've asked the right questions. Don't expect your manager to be able to answer everything at once; you might need to gather this information over the course of several meetings.

 Note

Importance of Notetaking

Don't underestimate the importance of taking detailed notes during this briefing. The volume of information shared can be overwhelming, and you'll rely on these notes as the foundation of your plan. Be prepared with a notepad or your preferred digital notetaking method to jot down critical points discussed during the meeting.

10.3.3 Step 3. Develop Your Plan

Flesh out the details in your plan template. Start with your notes from the previous step—this will get you off to a flying start. Working through each of the following chapters in Part 4: Plan in sequence will result in a comprehensive plan that sets your documentation project up for success.

Table 13: Topics in Documentation Plan Development

Topic	Description
Chapter 9: Collect Information	Explains how and where to collect existing information to inform both your plan and your approach to writing. Provides guidance on important setup steps such as securing access to systems.
Chapter 10: Make a Plan	Explains how to consolidate all the information gathered—including the details you will develop in the following steps—into an integrated management plan for your documentation project.
Chapter 11: Analyze Audience	Explains how to analyze your documentation's audience and break it down into primary, secondary, and hidden audiences so your documentation can be more effective.
Chapter 12: Define Review Team	Explains how to define a review team for documentation projects, an essential step in creating accurate and high-quality documentation.

 Tip

If you feel like you're ready to tackle more advanced project planning techniques, such as estimating and scheduling, follow the above steps in sequence. Then, check out Chapter 7: Estimate Scope, Time, and Cost, and Chapter 8: Develop Schedule in our forthcoming title, *Project Management for Technical Writers*, on our website: https://boffin.education. Use the coupon code in Templates to claim your one-year free subscription to the site.

10.3.4 Step 4. Refine Your Plan through Stakeholder Meetings

Set up a meeting or workshop with your stakeholders—the ones who have an interest in the documentation or those with whom you'll need to consult in the writing and review phases. Use this meeting to refine the plan until you're confident that it accurately captures everyone's expectations about the project. You may require several meetings to get it fully nailed down. Aim for brevity—there's no need for padding.

If you're documenting a product, don't forget to include the brand gatekeepers: product managers or the marketing team. Technical writers often tend to naturally affiliate themselves with the technical stakeholders, such as the engineering team, but it's important to recognize the role of the nontechnical players in creating a product as well. Technical writing isn't just about satisfying technical requirements; it's also a branding exercise.

>  **Tip**
>
> **Engaging Stakeholders in Documentation Planning**
>
> One of the hidden aspects of planning is the time required by subject matter experts and reviewers. Although it's not normally costed into a project, it's something you need to keep in mind because it can be a significant burden on the teams you're working with. Going through the steps of negotiating and agreeing to a plan ensures that the teams you're going to work with will be aware of—and hopefully supportive of—your future requests.

10.3.5 Step 5. Secure Approval for Your Plan

Before you proceed to the execution phase of your documentation project, it's essential to get your plan officially approved by your manager and key stakeholders. Approval is critical, as it serves as a testament that all parties have reached a mutual understanding of the project's scope, deliverables, and timelines. An approved plan acts as a protective measure and offers a reference point in case of future misunderstandings or disagreements regarding the project.

Additionally, obtaining a sign-off can re-engage stakeholders who may have been preoccupied, giving them the push that's needed to refocus so they can finally commit to the project officially.

Steps for Securing Approval:

1. **Prepare a Final Draft**: Ensure that your plan is free from errors and ambiguities.
2. **Circulate for Feedback**: Share the draft with stakeholders for any last-minute suggestions or corrections.
3. **Request Formal Approval**: Ask stakeholders for their official sign-off on the document. This could be a physical signature, an email confirmation, or an approval via a project management tool.
4. **File the Approved Plan**: Keep a copy of the signed plan in an easily accessible location, both for your reference and for audit purposes.

By following these guidelines, you can transition smoothly from the planning phase to the design, writing, and reviewing phases, with the confidence that everyone is aligned on the objectives and requirements.

Chapter 11 Analyze Audience

Lead Writer: Kieran Morgan | **Peer Reviewer/s**: Steve Moss

This chapter focuses on audience analysis in technical writing. We introduce the concept of audience categorization, grouping audiences into primary, secondary, and hidden groups, as we highlight their distinct needs. The Five Ws and One H method is introduced as a systematic approach for in-depth audience understanding. We also discuss the development of personas, drawing on techniques in customer experience design. The chapter guides readers through the practical steps of audience analysis, which sets the stage for developing user-centered technical documentation.

Who Should Read This

- Aspiring Technical Writers
- Beginner Technical Writers
- Career Advancers
- Cross-Domain Professionals

CONTENTS

11.1 Introduction	181
11.2 [Theory] Audience Types	182
11.3 [Theory] The Five Ws and One H	183
11.4 [Theory] Personas	187
11.5 [Practice] Analyze and Define Audience	189

Chapter 11 Analyze Audience

PROCESS

Inputs	Plan > Analyze Audience		Outputs
• Existing Technical Documents • Journey Maps and Personas • End-User Interviews • Subject Matter Expert Interviews	**Theory**	• Audience Types • The Five Ws and One H • Personas	• Audience Profile / Personas
	Practice	• Collect Information • Talk with Subject Matter Experts • Create Draft Personas • Conduct Audience Interviews • Analyze Data • Consolidate Audience Insights • Finalize Audience Profile and Validate	
	Examples	• Technical Writer Personas in Chapter 2: Technical Writing Roles and Responsibilities	
	Templates	• Audience Persona Template	

11.1 Introduction

"Having a lot of empathy is the most important skill for a tech writer. If you don't care about your users or the people you're working with there's going to be a disconnect eventually. Your writing will be ok—but it won't be exceptional."

—Dina, Head of Product Content and Knowledge

Imagine you're crafting a message—not just any message, but one that could transform how someone interacts with technology. That's the power of technical writing, which requires a fusion of empathy and expertise. In this chapter, we explore the heart of technical writing: understanding your audience. Remember the PADRE method we explored in Chapter 10: Make a Plan? In this chapter we bring that into play as we focus on your audiences' unique needs and perspectives.

Often technical writers find themselves on the sidelines, constrained by time or detached from project planning. Don't let this deter you from developing a thorough understanding of your audience. It's vital for any well-crafted document.

In this chapter, we're going to walk you through the process of categorizing audiences into three distinct types: primary, secondary, and hidden. We'll then unpack a classic technique known as the Five Ws and One H, a staple among writers across various fields. This method involves asking a set of crucial questions that help ensure you've considered your audience from multiple perspectives. We'll also explain persona development, drawing on techniques from the world of customer experience design. By doing so, you can create documentation that's not only informative but also highly relevant to your target audiences.

Finally, we'll lead you through the practical steps of audience analysis. You can put these concepts and steps into action in the audience profile in your Documentation Plan Template or take it a step further by using the Audience Persona Template to personify your audiences as archetypes.

With these tools, you'll gain a comprehensive insight into your audiences by building a foundation for technical writing that both informs and connects with its users.

> **What Does That Mean?**
>
> **Audience**
>
> - **Meaning 1**: ("An audience") A specific group or groups of people who will interact with or use a document. This can be categorized into primary, secondary, and hidden audiences, depending on their degree of interaction with the document.
>
> - **Meaning 2**: ("Audiences") A general term for anyone who interacts with or uses a document, regardless of their categorization.

11.2 [Theory] Audience Types

Before diving into the nuances of audience profiling with the techniques detailed below, it's best to categorize them into broad-brush categories. This helps you figure out which audience to prioritize in your documentation—and uncovers any audiences you might not have initially thought about.

Audiences can be categorized as primary, secondary, or hidden:

- **Primary Audience**: These are the direct recipients of your communication. For instance, "Our primary audience includes surgeons with varying levels of experience in using intraoperative devices."

- **Secondary Audience**: These audiences aren't the main focus but are still important. For example, "Our secondary audience includes nurses who assist in surgical procedures."

- **Hidden Audience**: These are audiences outside of the primary and secondary categories, often with common interests or who are indirect recipients. For example, "Our hidden audience encompasses medical equipment technicians who maintain and troubleshoot the intraoperative devices."

> **Note**
>
> **How Many Audiences Can There Be?**
>
> There may be multiples of any audience category, particularly secondary and hidden. So, how many should you define? There's no hard-and-fast rule on this, but it's helpful to remember the Pareto principle. This is the concept that roughly 80% of effects (or use cases for your documents) come from 20% of the causes (or audiences). So you don't need to capture every audience—but you do need to define the ones that count.

11.3 [Theory] The Five Ws and One H

The Five Ws and One H technique is a time-honored method of rhetorical analysis. Its origins are so old-school that they trace back to Aristotle. This approach has guided analysis from antiquity right through to modern times. In fact, for many years, it was a fundamental part of journalism training to ensure that news articles captured all the critical elements of a story in an objective and unbiased manner.[23]

It's an incredibly useful tool in audience analysis, and writing in general, which is why we've chosen it as our key theory for this chapter. While there are newer tools that we also explore in this chapter, such as personas, this one has stood the test of time, making it our first pick.

The Five Ws stand for Who, What, When, Where, and Why. The H is for How.

To use this technique for audience analysis, consider the following questions:

- **Who** is the audience for the documentation?
- **Why** does the audience need the documentation?
- **When** will the audience use the documentation?
- **Where** and **how** will the audience use the documentation?
- **What** is important to the audience?

- **What** challenges can the documentation help the audience overcome?

We've included the Five Ws question-and-answer format in our Documentation Plan Template as an example of how to apply this method. This is shown in the extract in the table below. Remember, you can use this technique to examine almost any situation!

Table 14: Example Audience Analysis using the Five Ws Technique for a Fictional Surgical Device User Guide

Question	Answer
Who is the audience? Key Question: Who will be reading or using this documentation?	Example Primary Audience: "Surgeons with varying levels of experience in using intraoperative devices." Example Secondary Audience: "Nurses who assist in surgical procedures." Example Hidden Audience: "Medical equipment technicians who maintain and troubleshoot the intraoperative devices."
Why does the audience need the documentation? Key Question: What problem or need does the documentation solve for the audience?	Example: "The audience needs the documentation to operate an intraoperative device effectively and safely because they are responsible for patient outcomes during surgical procedures."
When will the audience use the documentation? Key Question: At what stage in the user journey will the documentation be most useful?	Example: "The audience will use the documentation primarily during surgical procedures, and they aim to ensure both safety and efficacy."

11.3 [Theory] The Five Ws and One H

Question	Answer
Where and how will the audience use the documentation? Key Question: In what setting and through what means will the audience access the documentation?	Example: "The audience will use the documentation in the operating theatre while performing surgeries. They may also consult digital versions during pre-operative preparation."
What's important to the audience? Key Question: What are the top priorities or focus areas that the audience cares most about?	Example: "Accurate, step-by-step operational instructions are a top priority for the audience because any error could have severe implications for patient safety."

Chapter 11 Analyze Audience

Question	Answer
What challenges can the documentation help the audience with? Key Question: What are the specific challenges that the documentation aims to address for the audience?	Example Challenges for Primary Audience (Surgeons): • Ensuring surgical precision using new technology. • Balancing speed and safety during operations. Example Challenges for Secondary Audience (Nurses): • Quickly preparing and setting up new equipment. • Keeping pace with varied surgeon preferences and procedures. Example Challenges for Hidden Audience (Medical Equipment Technicians): • Understanding complex device operations for effective maintenance. • Diagnosing and resolving technical issues swiftly to minimize operation room downtime.

Note

The Ancient Roots of the Five Ws and One H

In Chapter 10: Make a Plan, we discussed the PADRE technique, an evolution of an ancient rhetorical method dating back to the era of the ancient Greek philosophers. The Five Ws technique also traces its origins to Aristotle, our philosopher friend who articulated the concept of rhetoric so eloquently.[24]

What Does That Mean?

Rhetoric

Rhetoric is the art of effective communication and persuasion. It involves connecting with your audience in a way that encourages them to view things from the speaker's or writer's perspective.

11.4 [Theory] Personas

A persona is a method of characterizing audiences that makes them easy to relate to and memorable. It's a commonly used tool in customer experience design. Personas, or archetypes, are fictional individuals representing the common characteristics of a group of customers.[25] In the case of technical documentation, this means your primary, secondary, or hidden audiences.

A persona builds on the audience analysis you've already done. You can weave your Five Ws analysis into your persona. But here's where it gets more interesting: a persona adds depth to this picture by getting a name, a visual avatar that represents them, and even a quotable quote they might say if they were a real person. This makes them both more memorable and more relatable. When folks start referring to your personas by their names, you'll know you've really struck a chord.

Chapter 11 Analyze Audience

Do you remember the technical writer types illustrated in the infographics in Chapter 2: Technical Writing Roles and Responsibilities? Those are examples of personas. We've used them to embody the different archetypes of technical writers. This was achieved by interviewing many technical writers, which allowed us to gain a comprehensive understanding of their professional lives, their thought processes, and the joys and challenges they experience.

To assist you in developing your own personas, we've put together an Audience Persona Template you can customize for your project.

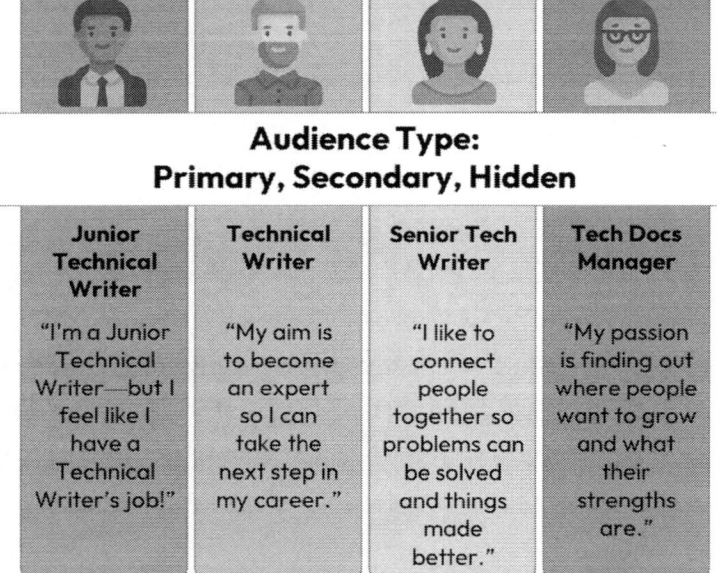

Figure 25: Example of Personas and Audience Types

Tip

Harnessing Personas for Consistent Writing

Developing personas is an optional yet powerful tool for guiding your writing team. It fosters consistency in how your documentation addresses and engages with its audience. This approach helps ensure that your technical content resonates more effectively with its intended readers.

 What Does That Mean?

Personas

A fictional character that embodies the common traits of a specific group of customers or audiences for documentation. These are often referred to as archetypes.

11.5 [Practice] Analyze and Define Audience

Use the steps below to analyze and define the audience for your documentation. We've drawn inspiration from Kalbach's five-step method, featured in his popular book on customer experience design, *Mapping Experiences*.[26]

11.5.1 Step 1: Collect Information

Begin by collecting available information as a resource. It's likely that existing technical documentation already defines the audience. Additionally, your organization may have customer journey maps with established personas that will give you a starting point.

Tip

Aligning Personas with Customer Journey Maps

When integrating personas into your documentation strategy, it's essential to align them with your organization's existing customer journey maps, if available. This ensures that the personas accurately mirror real-world customer experiences and needs. It also helps to ensure that these personas are consistent with your organization's existing research, further enhancing their relevance.

11.5.2 Step 2: Talk with Subject Matter Experts

Next you'll consult with subject matter experts, such as product managers and engineering team members. They're often well-versed in the audience's profile and needs. It's likely they have undertaken similar profiling activities, which will provide a valuable foundation for your own audience analysis. Keep your notepad ready—you'll want to capture as much detail as you can.

11.5.3 Step 3: Create Draft Personas

Use the templates provided to create a draft of your audience analysis and personas if you find them helpful. The templates in Chapter 10: Make a Plan and the Audience Persona Template are a great place to start. Think of these as your "straw person" audience personas—they might be reasonably accurate, but you won't know for sure until they're validated.

11.5.4 Step 4: Conduct Audience Interviews

If it's feasible (and if your organization's policies permit), validate your audience profiling by interviewing someone from your target audience or someone with similar characteristics. Aim to really step into these people's shoes as much as you can. This helps you understand the challenges they face when using the product, process, or technology.

If you can't interview your audience directly, chat with close proxies—ideally frontline folks like customer support who interact directly with your audience. If that's not possible, turn to people in your organization or your network who share similar characteristics, such as age, education levels, and abilities. Guide your questions with the Five Ws approach.

11.5.5 Step 5: Analyze Data

Now it's time to dive into your findings and pick out the common themes. A convenient method is to pull out key themes from each interview. Start by sorting them under categories like the Five Ws and One H. Then, get them organized into common themes. You can use a virtual whiteboard, such as Miro, for this. Just drag and drop related themes close to each other, and you'll soon see patterns emerging.

Use what you learn from this analysis to refine and confirm the initial straw person profiles you put together in the previous step.

11.5.6 Step 6: Consolidate Audience Insights

Now that you've gathered the information you need, it's time to bring it all together. Consolidate the insights from your interviews, SME discussions, and any existing documentation or customer journey maps. This step is about creating a comprehensive understanding of your audience's characteristics, needs, and preferences.

Begin by summarizing key findings and insights. Look for patterns and commonalities that help define the broader audience for your documentation. This might include technical expertise levels, common questions or concerns, preferred learning styles, and any other relevant factors. The goal here is not to create detailed personas but to form a clear, general profile of your audience.

11.5.7 Step 7: Finalize Audience Profile and Validate

With your consolidated insights, you can now finalize the audience profile. It should be a succinct, accurate representation of the people who will use your documentation. Include demographic information, common challenges, goals, and any other relevant details that emerged from your analysis.

Once the profile is drafted, it's important to validate it. Share the profile with your team and other stakeholders for feedback. Ensure it aligns with their understanding and experiences with the audience. Adjust the profile as needed based on their input.

Finally, make sure this audience profile is readily available to everyone involved in the documentation process. Consider adding it to your internal documentation resources, like a wiki or shared drive. A well-understood audience profile is a key tool in creating effective, user-centered technical documentation.

Chapter 12 Define Review Team

Lead Writer: Kieran Morgan | **Peer Reviewer/s**: Steve Moss

This chapter details the formation of a review team for technical documents. It highlights the importance of choosing the right mix of roles, such as approvers, editors, and subject matter experts. The chapter guides readers on organizing review processes and stakeholder involvement, featuring examples of review and approval matrices. Defining a review team early in the planning phase is essential for ensuring the accuracy, relevance, and quality of technical documents.

 Who Should Read This

- Aspiring Technical Writers
- Beginner Technical Writers
- Career Advancers
- Cross-Domain Professionals

CONTENTS

12.1 Introduction	195
12.2 [Theory] Who's Who in a Review?	196
12.3 [Practice] Define Review Team	199
12.4 [Sample] Sample Review Matrix	203
12.5 [Sample] Sample Approval Matrix	204

Chapter 12 Define Review Team

PROCESS

Inputs	Plan > Define Review Team		Outputs
• Stakeholder Conversations	**Theory**	• Who's Who in a Review	• Review Matrix in Documentation Plan
	Practice	• Identify Subject Matter Experts and Stakeholders • Classify Reviewers and Approvers • Identify Lead Times and Sequence • Finalize Review Matrix	
	Examples	• Sample Review Matrix • Sample Approval Matrix	
	Templates	• Review Matrix in Documentation Plan Template	

12.1 Introduction

"It's frustrating when you can't get someone to review your docs because they're too busy. But if you give them a little nudge then they'll usually do it. Don't take it personally."

—*Francis, Technical Writer*

Ever felt the frustration of hanging in limbo, waiting for someone else to play their part? You're not alone. Many of the writers we interviewed share this common plight. In technical writing, it's wise to establish your review team at the outset of your project. The review phase often introduces an element of unpredictability. You might receive unexpected feedback leading to further revisions or come up against the challenge of involving overloaded stakeholders. Such hurdles can sidetrack even the most skilled writers. Many professionals we spoke with echoed this sentiment, citing review-related obstacles as a significant source of frustration.

Selecting the right review team is essential. This step ensures your work is not just accurate and high quality, but also resonates with the expectations of your stakeholders. Formal reviews serve as a testament to the rigor and governance your documents have undergone. These can vary from straightforward email approvals to more complex compliance with organizational document control policies.

This chapter will guide you in defining a review team as we highlight the typical roles and stakeholders involved in document review. We'll also present an example of a well-structured review matrix. Defining the review team in the early stages of planning can significantly enhance the likelihood of your project's success.

> **What Does That Mean?**
>
> **Review**
>
> The process of evaluating a document against quality standards such as technical accuracy, consistency with style manuals, templates, branding, and so on.
>
> **Approval**
>
> The formal acknowledgment or sign-off by an approver or document owner that a document is fit for publication.

12.2 [Theory] Who's Who in a Review?

12.2.1 Generic Roles in a Review Team

The table below outlines the generic role names for participants in a document review. These role names represent broad categories of reviewers rather than specific job titles like engineering, product management, and so on.

Table 15: Generic Roles in a Documentation Review Team

Role	Definition
Approver	An authority with decision-making power on whether to endorse a document for the next phase, usually publication.
Document Owner	Holds responsibility for documents within a specific topic, like electrical engineering, software products, and so on. Usually also an approver.

12.2 [Theory] Who's Who in a Review?

Role	Definition
Editor	Responsible for editing and proofreading the document. This role could be filled by the writer or someone more specialized.
Moderator / Curator	Acts as a gatekeeper reviewing content against organizational standards before publication.
Peer Reviewer	Reviews for adherence to writing and design principles, as well as organizational, team, local, or international standards. They often have substantial general knowledge of the organization and bring significant value to the review process. Peer reviewers are typically members of the technical documentation team or function.
Reviewer	A subject matter expert or stakeholder who uses their expertise to improve and validate the quality of a document through a review process.
Subject Matter Expert	An expert with deep domain knowledge in an area covered in the document, such as technical, marketing, product management, legal, and so on.
Writer	Responsible for orchestrating and executing the planning, designing, writing, review, approval, and publication of a document. Responsible for initiating review and approval requests.

12.2.2 Typical Stakeholders in a Document Review

This table outlines the usual stakeholders typically involved in a document review. It's important to note that the precise makeup of the review team will always vary from one project to another. It's not a "one size fits all" approach—it's something you need to tailor for each project and document.

Table 16: Typical Stakeholders in a Documentation Review

Stakeholder	Responsibility
Engineers / Developers	Ensuring technical accuracy and alignment with specifications. Verifying the accuracy of functionality and user interface as depicted in the documents. Making sure the content is appropriate for the intended audience/end users.
Intellectual Property	Proper acknowledgment of trademarks, including third-party ones. Correct assertion of copyright, trademarks, and other intellectual property rights.
Legal / Regulatory	Adherence to relevant regulations, laws, and standards, which may vary depending on the organization, industry, and geographical location. Ensuring the legal terms and conditions are accurate.
Marketing / Communications	Staying in line with company brand guidelines and the style guide. Using the correct branding elements (such as templates, logos, colors).
Process Owner	Verifying the accuracy of process flows in process-based documentation.

Stakeholder	Responsibility
Product Managers	Ensuring alignment with product branding and naming conventions. Verifying the accuracy of the representation of product functionality and user interface in the documents. Assessing the suitability of the content for the audience/end users.
Project Manager / Project Sponsor	Ensuring alignment with project objectives and deadlines.
Quality Manager / Process Manager	Compliance with management system standards and templates.
Risk Management	Documenting any health and safety risks inherent in the product (such as sharp edges, small parts). Ensuring safety warnings are appropriately worded and displayed.
Writer's Line Manager	Mentoring and coaching of writers through the provision of feedback. Quality check against expectations. Ensuring alignment with the writing team's standards and style. Usage of the correct template.

12.3 [Practice] Define Review Team

Defining a review team is straightforward. Just follow the steps below, and use the review team section in your documentation plan to capture all the details.

> **Tip**
>
> **Do Your Detective Work Before Diving In**
>
> Defining a review team is a voyage of discovery, especially if you're new to an organization. Instead of nominating who does what, it's a process of investigating and discovering who the appropriate authorities are. In some organizations—particularly government departments—there are well-defined hierarchies of managers and strict procedures that you must follow. In others, such as those using Agile methodologies, this formality might go against the grain! It's important to take the time to understand your organization's culture before diving in.

12.3.1 Step 1: Identify Subject Matter Experts and Stakeholders

In the planning stages of your writing project, consult with your manager or project leader. Ask about who the subject matter experts (SMEs) and stakeholders are for your documentation. It's key to understand their areas of expertise—for instance, whether they focus on product development or engineering—and their specific specializations within these fields. Figure out their priorities and check if they have any unique requirements or preferences, like needing a longer notice period for reviews.

Insight

Why Define a Review Team in the Planning Phase?

Might it seem easier to wait until you've finished writing before tackling the review process? Actually, no. Defining your review team in advance helps you avoid the chaos of a poorly planned review and the havoc that can wreak on your project timeline.

Doing so enables you to:

- Provide busy subject matter experts with adequate notice about their roles as reviewers or approvers.
- Guide subject matter experts to offer the most useful feedback, such as comments on technical accuracy, rather than critiques of your grammar and punctuation.
- Establish a framework for the scope of reviews, ensuring everyone's time is well-utilized. For example, you can ask legal reviewers to focus on legal and compliance issues instead of commenting on technical content.
- Clarify your responsibilities as the writer. This makes it clear to everyone what you handle and highlights the value your expertise adds in managing the process.
- Identify gatekeepers early in the process, ensuring that your review and approval workflow adheres to any specific protocols required.

12.3.2 Step 2: Classify Reviewers and Approvers

Identify who in your group needs to review the document and who the approvers are. Use the Review Matrix found in your documentation plan to categorize them accordingly.

Keep in mind, some stakeholders might take on dual roles as both reviewers and approvers. Typically, more senior stakeholders act as approvers, leaving the detailed document review to their less senior colleagues, who often have more time to dedicate to this task.

12.3.3 Step 3: Identify Lead Times and Sequence

This step is critical: you need to identify the lead times for your review team. Though document reviews might only take an hour or two, people usually need at least a few days' notice to fit them into their busy schedules.

Also check if any parts of the review and approval process have to be done in sequence, like moving from junior to senior approvers. This is common in traditionally structured organizations, like government agencies, and can significantly extend your timeline.

 Tip

Identify Gatekeeping in Advance

It's essential to identify any gatekeeping well in advance. Some teams, especially legal teams, are constantly inundated with document reviews and may only respond to formal requests. This could mean sending a document to a team inbox or submitting a request via a form on an intranet site. Teams like these often have lengthy lead times they strictly adhere to and might not respond to requests sent through informal channels.

12.3.4 Step 4: Finalize Review Matrix

Finally, incorporate all this information into the review matrix in your documentation plan and review it for completeness. You might need to update this section as you progress. Sometimes you'll find that the person nominated wasn't the ideal choice or that they've moved on to another role or project.

12.4 [Sample] Sample Review Matrix

We've put together a sample review matrix below to give you a feel for how yours might look in your documentation plan. This example is set in a fictional scenario where a technical writer is crafting medical documentation.

Table 17: Sample Review Matrix

Team	Name and Job Title	Contact Info	Review Focus Area	Lead Time (Business Days)
Surgical Team	Dr. Jane Smith, Lead Surgeon	jsmith@example.com	Surgical procedures and medical terminology	5
Nursing Team	Emily Johnson, Nurse Coordinator	ejohnson@example.com	Equipment setup and nursing procedures	3
Legal	Laura Williams, Legal Advisor	lwilliams@example.com	Regulatory compliance	7

Team	Name and Job Title	Contact Info	Review Focus Area	Lead Time (Business Days)
Tech Team	Mark Lee, Software Engineer	mlee@example.com	Software interface and connectivity	4

12.5 [Sample] Sample Approval Matrix

See the table below for what an approval matrix could look like in the same fictional project. Notice how the approval process is sequential in this example. While this approach ensures thorough review, it can significantly extend the overall project timeline.

Table 18: Sample Approval Matrix

Team	Name and Job Title	Contact Info	Approval Sequence (#)	Lead Time (Business Days)
Quality Team	Sarah Brown, Quality Manager	sbrown@example.com	1	6
Medical Board	Dr. Alan Watts, Board Chairman	awatts@example.com	2	10
Legal	Laura Williams, Legal Advisor	lwilliams@example.com	3	7

Part 5 Design

Harnessing design principles in technical writing to structure documents, design visual stylesheets, and create document templates.

Chapter 13 Design Structure

Lead Writer: Kieran Morgan | **Peer Reviewer/s**: John New | **Expert Reviewer/s**: Dina Bennett, Patrick Lambe

In this chapter, you'll explore the art of organizing technical documents to make them more user friendly and effective. You'll learn the importance of structure, the factors that differentiate it from information architecture, and the common taxonomies—lists, tree structures, process flows, system maps, and matrices—used in technical writing. Real-world examples and steps guide you through the process of building and iterating your document's structure to ensure it aligns with your audience's needs.

Who Should Read This

- Aspiring Technical Writers
- Beginner Technical Writers
- Career Advancers
- Cross-Domain Professionals

CONTENTS

13.1 Introduction	209
13.2 [Theory] The Content Sandwich: Front Matter, Text, and Back Matter	210
13.3 [Theory] The Five Taxonomies	212
13.4 [Theory] Sequential and Nonsequential Taxonomies	219
13.5 [Practice] Design Structure	221
13.6 [Practice] Review and Iterate Structure	222
13.7 [Case Study] Cochlear™ Nucleus® 6 Sound Processor User Guide	223

Chapter 13 Design Structure

PROCESS

Inputs	Structure > Design Structure		Outputs
• Documentation Plan • Consolidated Background Information	**Theory**	• The Content Sandwich: Front Matter, Text, Back Matter • The Five Taxonomies • Sequential and Nonsequential Taxonomies	• Document Structure (e.g., Table of Contents)
	Practice	• Design Structure • Review and Iterate Structure	
	Examples	• Cochlear™ Nucleus® 6 Sound Processor User Guide	
	Templates	• Sample User Guide	

13.1 Introduction

"Taking the clay of what the developers are working with and building it into bricks is the most important part of the job."

—*Colin, Senior Technical Writer*

Technical writers love to build structure. There's nothing more satisfying than taking a bunch of scattered facts and synthesizing them into a well-organized technical document. Many of the technical writers we interviewed for this book reported that this was one of the main joys of their working lives. It's like fitting a puzzle piece into place and suddenly seeing the bigger picture come into view.

It's not just satisfying for us, the writers; it also makes our documents more effective for our audience. When we organize information effectively, we make it easier for our users to locate the information they need quickly. This helps them achieve their goals faster—whether it's building knowledge about how to use something or obtaining the information they need to make a decision.

When we think about structure, the thing that comes to mind most immediately is the hierarchy of headings that forms a document's outline, otherwise known as a table of contents. But structure goes far deeper than that. We're going to introduce you to several concepts in this chapter, followed by steps to apply them in practice.

The first is what we call the "content sandwich"—it's an old-fashioned but still useful means of understanding the structure of long-form documents, such as books, reports, plans, and printed documents. This method is still commonly used in publishing today.

The second is an introduction to what we call the Five Taxonomies: the most common types of structures used in technical documents. These are lists, tree structures, process flows, system maps, and matrices. You've probably used these already but maybe without realizing they had specific terminology.

The third is the distinction between sequential and nonsequential taxonomies. This is the difference between instructions that must be followed in a particular order versus content that doesn't require a particular sequence to grasp it.

Understanding how to apply the right taxonomy will help you create a user-friendly document that best supports your audience's journey.

13.1.1 Structure vs. Information Architecture

In your technical writing career, you'll probably hear the words *structure* and *information architecture* used interchangeably. There's a lot of overlap between these concepts and no single, agreed-upon definition—so if you're a bit confused, you're not alone.

Here's how we differentiate them in this book:

- **Structure** refers to the organization within a technical document, such as a table of contents, list of procedures, or flowchart.
- **Information architecture** is a holistic system of information design that encompasses both the internal structure of a document and the labeling, navigation, and search subsystems beyond the document.[27]

While we would love to explore information architecture, it's beyond the scope of this book.

13.2 [Theory] The Content Sandwich: Front Matter, Text, and Back Matter

Once upon a time, the design and structure of documents were guided by well-understood conventions that had been in place for decades, if not centuries. Content is traditionally divided into several components: front matter, text, and back matter. Think of this structure like a content sandwich, with the front and back matter serving as the bread and the text as the tasty filling.

13.2 [Theory] The Content Sandwich: Front Matter, Text, and Back Matter

Here's a breakdown of these elements:

- **Front Matter**: Typically contains the title page, version history, copyright statement, and navigational aids such as the table of contents.
- **Text**: Comprises the "nutritional value"—the valuable information about the topics and tasks that the user wants to accomplish or learn about.
- **Back Matter**: Includes all other content not essential to the topic, such as appendices and glossaries. Often incorporates additional navigational aids like indexes and tables of figures.

Finally, there's an outer layer, or "wrapper":

- **Cover**: This is where the entire content sandwich is wrapped up, as if in a metaphorical paper bag. The cover serves as the final layer, providing a visually attractive design to distinguish it from other books and to protect the hard copy from damage.

If you consult your nearest style guide, you'll invariably find a section offering detailed guidance on laying out traditional publishing deliverables, such as books and journals. While this information is valuable, it's primarily targeted at the publishing industry. So is it really relevant to you? Yes! Understanding these conventional structures can assist you in organizing your technical documents more effectively. It allows you to compartmentalize the components of your document, enabling more manageable planning, especially if you're designing something lengthy—such as a report, plan, or any document that will be printed and bound as a hard copy.

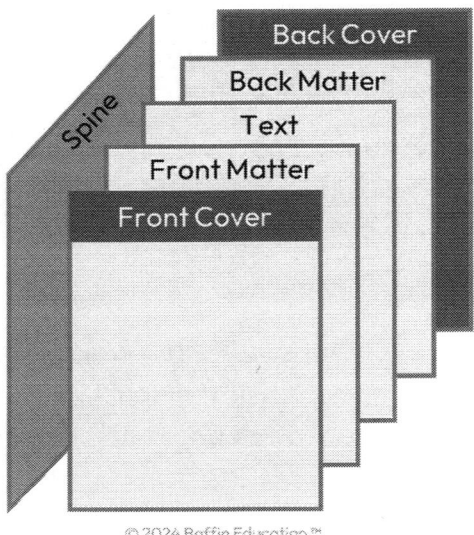

Figure 26: The "Content Sandwich": Front Matter, Text, Back Matter, and Cover

 What Does That Mean?

Front Matter, Text, and Back Matter

In publishing, content—such as a book—is traditionally divided into several sections: front matter, text, back matter, and a cover. This concept is not just academic—it also applies to many other long-form or hard-copy (i.e., printed and bound) documents, such as reports, plans, and user guides.[28]

13.3 [Theory] The Five Taxonomies

Let's say you're at the grocery store standing next to a shelf full of oranges. On the next shelf, you find lemons. Where are you? You're in the fruit section, of course. What's nearby? Most likely the vegetable section.

13.3 [Theory] The Five Taxonomies

How did you know that without even thinking about it? Because it's a taxonomy—a powerful tool that helps us make sense of the world by organizing knowledge in our minds and aiding us in decision-making.[29] The taxonomy at your grocery store allows you to navigate it intuitively, as the store designers have purposely co-located vegetables and fruit to mimic the way we've learned since childhood to mentally classify food.

As a technical writer, you create taxonomies all the time. A table of contents in a user guide, a sequence of ordered steps in a simple procedure, a document describing a business process flow, a matrix showing software version and compatibility—all of these things are taxonomies.

Good taxonomies make technical information predictable and easy to navigate by mimicking the process that occurs inside the human brain. If you understand the common types of taxonomy, you can structure your information in predictable ways that will seem familiar to your audience, just like a shopper in a grocery store.

The five most common taxonomies in technical documentation are lists, tree structures, process flows, system maps, and matrices:

The Five Taxonomies in Technical Documentation

	LIST	TREE STRUCTURE	PROCESS FLOW	SYSTEM MAP	MATRIX
DESCRIPTION	A collection of items grouped by a common purpose or function, without a hierarchical structure.	A branching structure with multiple levels.	A visual representation of sequential steps and decisions within a process.	Visual depiction of things that interact to create an integrated whole, or a system.	A two- (or more) dimensional table that allows for cross-referencing items.
EXAMPLES	• Shopping list • Sequential steps in a procedure • Simple index in a book (single-level)	• Table of contents in a book • Organizational chart • Index in a book (multi-level)	• Process flow diagram	• Network architecture diagram • Infographic of engine illustrating fluid dynamics	• Compatibility chart between software versions and devices • Risk assessment matrix

© 2024 Boffin Education™

Figure 27: The Five Taxonomies in Technical Documentation

13.3.1 Lists

A list is the simplest taxonomy. A list is a collection of items grouped by a common purpose or function, without a hierarchical structure. Lists are the building blocks of taxonomies.

- **When to Use**: Ideal for straightforward information that doesn't require branching or complex relationships. Useful in checklists, step-by-step instructions, or itemized data.

Example 1

A list of system requirements necessary to run software:

- Windows 10 or later
- 8GB RAM
- 100GB free disk space

Example 2

A guide to installing software, where each step is listed in a sequence to be followed by the user:

1. Download the installer.
2. Run installer.exe.
3. Follow the on-screen prompts.
4. Complete the installation.

13.3.2 Tree Structures

Tree structures are what we often think of when considering taxonomies. They are called "tree" structures because they start at a "trunk" and then expand into multiple "branches."

We subconsciously use tree structures all the time to make sense of our world; a classic example is a family tree. This makes them powerful tools for expressing relationships in our documentation—we all intuitively understand them, so there's no need to explain the concept of a tree structure to anyone.

You can use tree structures to represent numerous relationships, such as parent-to-child, container-to-contained, and cause-and-effect. Tree structures can also express horizontal relationships, such as siblings along one branch. This is akin to the co-location of fruits and vegetables in a grocery store, where items of similar nature are grouped together.

- **When to Use**: Best for organizing information in a branching format, where main topics are broken down into subtopics, like chapters in a book.

Examples

A table of contents in a user guide, showing main topics (parents) and subtopics (children):

1. About This Document

1.1. Overview

1.2. Revision History

2. Getting Started

2.1. In This Section

2.2. Log in

13.3.3 Process Flows

Process flow diagrams are commonly found—and highly useful—tools in the workforce. Most experienced technical writers have documented or created process flows at some point in their career, and there's even a type of technical writer that specializes in this work—see Chapter 2: Technical Writing Roles and Responsibilities: Types of Technical Writers.

Chapter 13 Design Structure

Process flows visually illustrate the steps, decisions, and interactions involved in a process. Process flows can usually be easily identified as they show direction in workflow, typically from an input to an output, along with the actions and decisions in between. This distinguishes them from system maps (below), which show relationships between components.

- **When to Use**: Effective for illustrating step-by-step processes, workflows, or procedures where a sequential flow is essential.

Examples

1. **User Onboarding Process**: A step-by-step guide to show what a new user needs to do to successfully set up an account and start using software.
    - Sign-up → Email Verification → Profile Setup → Tutorial → Full Access
2. **Software Deployment Workflow**: A diagram that shows the flow from code commit, to testing, to deployment in a production environment.
    - Commit → Automated Testing → Manual Review → Deployment

---

 What Does That Mean?

Process

A set of activities or tasks performed to accomplish an objective. Processes typically have a trigger that initiates the process; inputs necessary to perform the process; a corresponding result, or output; and a sequence of steps and decision points in between.

13.3.4 System Maps

System maps are visual representations of things that interact to create an integrated whole, or a system. These can be real-world or more conceptual in nature. They use a combination of text and visual symbolism to express relationships between items in their taxonomy. System maps are excellent tools for visualizing how the parts of complex systems interact with one another, making them a great choice for technical documentation.

- **When to Use**: Ideal for representing complex systems (physical or conceptual) with multiple interconnected components, where the interaction between components is key.

System Maps vs. Component Diagrams

In technical documentation, both system maps and annotated component diagrams are used to explain parts of a system. However, they have some key differences. While annotated diagrams label components, offering a static snapshot, a system map goes a step further to illustrate relationships, interactions, or flows between these components.

Examples

System Maps

1. **Jet Engine Schematic**: Schematic of a jet engine illustrating fluid dynamics.
2. **Network Architecture Map**: A diagram that visualizes how servers, routers, and firewalls are interconnected within an organization's network, including data flow directions.

Component Diagrams

3. **Laptop Hardware Overview**: A diagram that labels the main components of a laptop, such as the CPU, RAM, and ports, without detailing their interactions.

Chapter 13 Design Structure

> **What Does That Mean?**
>
> **System**
>
> An integrated whole consisting of interacting components. These components can be both tangible and intangible, such as the people, processes, knowledge, equipment, and software within a company.

13.3.5 Matrices

A matrix is a two- (or more) dimensional table that allows for the cross-referencing of items along rows and columns. In technical documentation, matrices are particularly useful for correlating data, considering different combinations of attributes, representing complex relationships, and aiding in decision-making.

- **When to Use**: Useful for comparing multiple variables or showing relationships between different sets of data, like in risk assessment matrices or skill matrices.

Example: Software Compatibility Matrix

	Windows 10	Windows 11	macOS
Version A	Compatible	Not Compatible	Compatible
Version B	Not Compatible	Compatible	Not Compatible

More complex matrices, such as three-dimensional data cubes, may have multiple dimensions, but the ones you'll typically find in documentation only have two dimensions. Matrices like this one make it easy to determine the compatibility between software versions and operating systems. This taxonomy type adds an additional layer of information that makes your documentation even more user friendly.

13.4 [Theory] Sequential and Nonsequential Taxonomies

So now you know the content sandwich and taxonomies. Let's explore another key concept in technical writing: sequential and nonsequential taxonomies. This distinction determines whether a document's content is arranged in a sequential order or not. Understanding when to use these structures will enable you to select the most appropriate taxonomy for your audience.

- **Sequential structures** resemble a series of instructions, like a recipe, where each step builds upon the previous one and must be followed in a precise order to achieve the desired outcome.
- **Nonsequential structures** are like the layout of an encyclopedia: individual pieces of information stand alone and can be accessed in any order.

Knowing when to apply each structure is an essential skill for technical writers. The following scenarios illustrate the practical use of sequential and nonsequential structures:

Structure	When to Use	Examples
Sequential	When accuracy depends on the completion of steps in a specific sequence	Procedures and process-based documentation Step-by-step how-to guides Task-oriented sections in user manuals

Structure	When to Use	Examples
Nonsequential	Where task execution does not depend on a particular order When organizing information by themes or features rather than a process	Knowledge base articles Frequently asked questions (FAQs) Feature descriptions in user guides Glossaries Product specifications

As a further example, let's consider the sections in our *Sample User Guide*:

- **Introduction**: Semi-sequential; this section provides an overview and can be read in any order, placing the most important information first.
- **Getting Started**: Sequential; this section guides the user through a set process.
- **Features and Functions**: Nonsequential; it describes product features independently of each other.
- **How-to Guides**: Sequential; these guides contain step-by-step procedures for common tasks.
- **Troubleshooting**: Nonsequential ordering of topics, with sequential ordering of steps within each topic.

This example demonstrates how the appropriate choice of sequential and nonsequential structures can create a user-friendly document. Each section's structure is selected to best support the user's journey through the information.

13.5 [Practice] Design Structure

The journey to a well-organized structure begins with your understanding of your audience's needs, your subject matter, the type of documentation you'll be crafting, and how it will be delivered to end users. When you're building your structure, you'll be using the foundation of understanding you developed in Part 4: Plan, particularly the audience analysis in your documentation plan.

Here's how you do it:

1. **Understand Your Audience**: Start by developing as deep an understanding as you can of your audience, the tasks they need to accomplish, and their information journey by following the steps in Chapter 11: Analyze Audience.

2. **Understand Your Subject**: Dive deep into your subject matter. Get as much hands-on experience as you can by following the steps in Chapter 9: Collect Information. There's no point documenting something you barely understand.

3. **Interview Subject Matter Experts**: Consult with subject matter experts to grasp the full scope of your documentation. Ask about key use cases, features, or modules. It might be necessary to conduct multiple interviews to draft a realistic structure or table of contents.

4. **Consult Stakeholders**: Discuss with stakeholders like your project sponsor or manager to understand their expectations. This is vital to refine your approach before finalizing anything.

5. **Use Your Intuition**: Once you've completed your planning and information gathering, a structure will likely begin to form intuitively. It may differ from your initial concept, but that's perfectly normal. Use this intuitive structure as a working draft to guide your further development.

13.6 [Practice] Review and Iterate Structure

Even the most meticulously thought-out structure will benefit from review. As with the writing and review process, this is an iterative step, meaning you might have to go back and forth multiple times. That's perfectly normal. By reviewing and iterating your document's structure before you commit to writing, you'll save yourself from having to rework your document later on and to bring your stakeholders along on the journey.

Here's our suggested process for finessing your document's structure:

1. **Expert Review**: Send your initial table of contents or structure draft to subject matter experts and stakeholders.
2. **Gather Feedback**: Solicit feedback from your team and others, such as potential users. Ensure you've covered all the important topics and addressed the questions they might have.
3. **Revise**: Make changes based on the feedback you've received.
4. **Final Review**: Once you're satisfied with your structure, it's time to start writing.

By developing an understanding of the different types of document structure and following the steps above, you can ensure you're creating a well-organized document that's better suited for your audience's needs. When you feel like your draft structure is sufficiently well-developed, it's time to start writing.

13.7 [Case Study] Cochlear™ Nucleus® 6 Sound Processor User Guide

A great example of structure in action can be found in Cochlear's range of sound processors for hearing-impaired people. While developing the processor documentation for the Cochlear™ Nucleus® 6 Sound Processors, the technical documentation team used a combination of tree structures, component diagrams, and lists of procedures to create highly visual documentation suitable for its end users.

Kieran Morgan, who was Project Manager for the Cochlear™ Nucleus® 6 documentation, explains: "We put a lot of effort into the planning stage to really understand our audience's needs. By working with audiologists, engineers, and product managers, we realized our documentation had to cater to a wide range of demographics, including older adults and children worldwide with varying levels of literacy. So we developed a tree structure table of contents based on simple, sequential customer use cases like 'power,' 'turn on,' 'wear,' and 'care.' We color-coded these use cases and assigned them unique visual symbols to reinforce the taxonomy throughout the document, making it easier for even low-literacy customers to navigate the printed booklets and find the information they were looking for."

Chapter 13 Design Structure

Figure 28: Example Component Diagram in Cochlear™ Nucleus® 6 Sound Processor User Guide

13.7 [Case Study] Cochlear™ Nucleus® 6 Sound Processor User Guide

Contents

Power 5
Batteries 6
Battery life 6
Replace the battery 7
Lock/unlock the battery module to the processor 8
Lock/unlock the tamper resistant battery cover 9
Charge rechargeable battery modules 10
Change disposable batteries 11

Use 13
Turn on and off 14
Change programs 15
Pair with remotes 15
Put on your processor 16
Lock and unlock buttons 18
Wireless Accessories 19
Plug-in audio accessories 20
Telecoil 23

Wear 27
Sport and exercise 28
Wear a Mic Lock 29
Wear a Snugfit 30
Wear LiteWear 33
Personalise your processor 34
Travel 35

2

Figure 29: Example Color-Coded TOC in Cochlear™ Nucleus® 6 Sound Processor User Guide

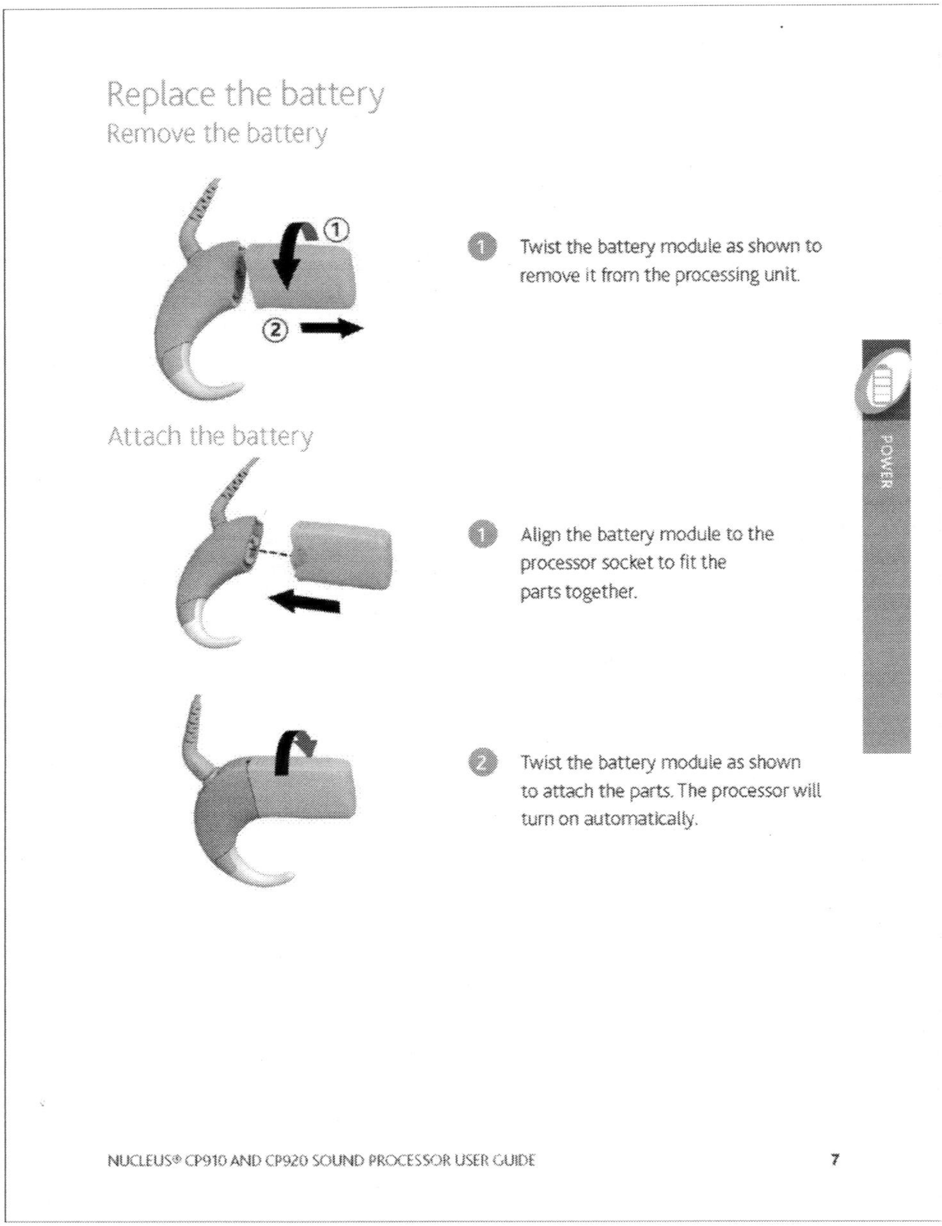

Replace the battery

Remove the battery

1. Twist the battery module as shown to remove it from the processing unit.

POWER

Attach the battery

1. Align the battery module to the processor socket to fit the parts together.

2. Twist the battery module as shown to attach the parts. The processor will turn on automatically.

NUCLEUS® CP910 AND CP920 SOUND PROCESSOR USER GUIDE 7

Figure 30: Example Page in Cochlear™ Nucleus® 6 Sound Processor User Guide Showing Navigational Aids and Component Diagrams.

Chapter 14 Design Stylesheet

Lead Writers: Swapnil Ogale, Kieran Morgan | **Peer Reviewer/s**: John New | **Expert Reviewer/s**: David Whitbread, Deirdre Wilson

This chapter focuses on the use of stylesheets in document design. It explores the theory behind stylesheets, explaining their role in separating presentation from content and structure. The chapter further explains key graphic design principles—contrast, alignment, repetition, and proximity (CARP)—and visual reading patterns. It provides practical examples to demonstrate the effectiveness of well-designed stylesheets. This chapter is essential for anyone looking to master the art of document design in technical writing.

 Who Should Read This

- Aspiring Technical Writers
- Beginner Technical Writers
- Cross-Domain Professionals

CONTENTS

14.1 Introduction	229
14.2 [Theory] Separation of Presentation, Content, and Structure	230
14.3 [Theory] Stylesheets	232
14.4 [Theory] Contrast, Alignment, Repetition, Proximity (CARP)	234
14.5 [Theory] Visual Patterns of Reading	242
14.6 [Practice] Design Stylesheet	244
14.7 [Sample] Sample Business Card Stylesheet	245

Chapter 14 Design Stylesheet

PROCESS

Inputs	Design > Design Stylesheet		Outputs
• Brand Guidelines	**Theory**	• Separation of Presentation, Content, and Structure • Stylesheets • Contrast, Alignment, Repetition, Proximity (CARP) • Visual Patterns of Reading	• Document Stylesheet
	Practice	• Design Stylesheet	
	Examples	• Sample Business Card Stylesheet	
	Templates	• Visual Design Principles Checklist for Technical Writers	

14.1 Introduction

In this chapter, we introduce you to the theory and practice of designing a stylesheet for your documents. A stylesheet governs the visual styling—the fonts, colors, spacing, and so on—that you immediately see when you open a document.

Have you ever flicked through a book of house designs, finding yourself drawn to one and hating another, purely based on aesthetics such as color and pattern? This is like the "presentation" layer in a document—it's the visual rendition that we all see when we walk into a house: the color of the paint on the walls, the choice of decorations, and so on.

Oftentimes, you'll find that two houses with the same floorplan look completely different. Perhaps one has a traditionally light and elegant Hamptons-style facade, while another features a modern aesthetic with brushed steel and dark timbers and paints. This concept—the same underlying structure with multiple visual design options—is similar to a stylesheet for a document. Stylesheets set up rules for formatting the visual elements in your document, providing a convenient way of changing the presentation layout instantaneously. That's what makes stylesheets such powerful tools—they remove the need for laboriously updating every document any time a minor design change is made.

Imagine if you could do that with your house!

Designing a stylesheet is something every technical writer needs to understand. In this chapter, we explore several underpinning concepts. The first is the separation of presentation, content, and structure, which provides a layered view of a document, similar to our house metaphor above. The next is the graphic design principles that combine to make your document more visually effective—Contrast, Alignment, Repetition, and Proximity (CARP). Another is the visual patterns of reading that show how we tend to read on-screen in geometric patterns, reinforcing the importance of the alignment principle.

Finally, we'll give you some worked examples of these principles being applied in the creation of a simple stylesheet. This should give you a grasp of the fundamentals of the theory and practice of designing stylesheets.

14.2 [Theory] Separation of Presentation, Content, and Structure

The separation of presentation, content, and structure is a foundational concept in technical writing, publishing, and content management.[30] It's something every new technical writer needs to grasp, as it has numerous benefits, which we'll discuss shortly.

Think of presentation, content, and structure as layers, each building on and reinforcing the others. At the bottom is the content layer, which is overlaid by the successive layers of structure and presentation to create a whole document.

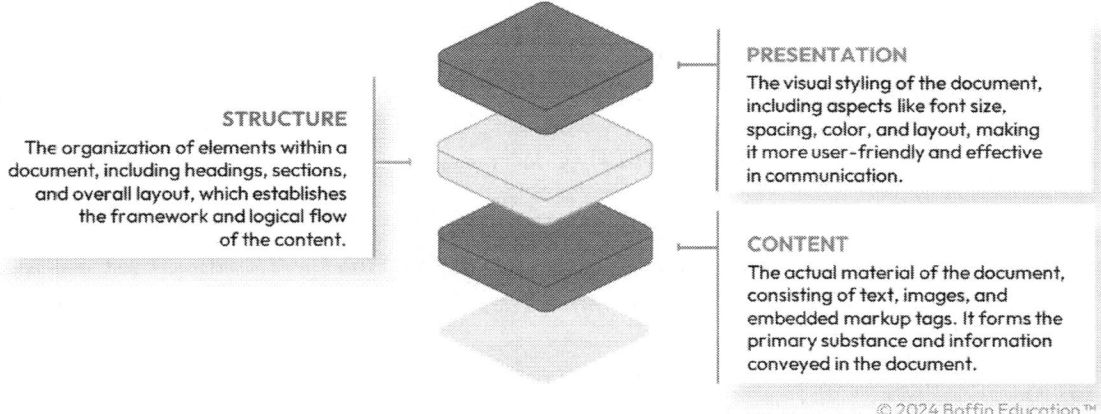

Figure 31: The Separation of Content, Presentation, and Structure

14.2 [Theory] Separation of Presentation, Content, and Structure

Here's how it breaks down:

- **Content**: First, let's consider the foundation: the actual text of your document—the words and images you put in there, and any semantic markup tags you apply to it, such as <p> (paragraph) or <h1> (heading). This is like the bricks, mortar, and timber that make up your house. The content layer contains the building blocks of meaning, which is why it forms the foundation.

- **Structure**: In Chapter 13: Design Structure, we learned that structure refers to the organization of information into taxonomies within a technical document, such as the headings that make up the table of contents. Think of document structure like a blueprint for building a house—similar to the floor plan you might see when browsing real estate listings. It organizes content in its correct place relative to other information.

- **Presentation**: Finally, the presentation layer adds a visual element—font sizes, spacing, colors, and so on—over the structure and content layers. Think of it like the paint job inside a house—it's akin to the interior design which appeals (or not) to your sense of style. Just like good interior design also improves the "usability" of a house, a good presentation layer makes your document more effective by making it easier for users to find the information they're after.

Chapter 14 Design Stylesheet

> **What Does That Mean?**
>
> **Content**
>
> The actual material of the document, consisting of text, images, and embedded markup tags.[31] It forms the primary substance and information conveyed in the document.
>
> **Structure**
>
> The organization of elements within a document, including headings, sections, and overall layout, which establishes the framework and logical flow of the content.
>
> **Presentation**
>
> The visual styling of the document, including aspects like font size, spacing, color, and layout, making it more user friendly and effective in communication.

14.3 [Theory] Stylesheets

In documentation, the content, structure, and presentation layers each complement one another. Just like you wouldn't build a home without a blueprint, good-quality materials, and consideration for interior design, you wouldn't design a document without considering the distinct but related layers of content, structure, and presentation, and how they work together to make your document a functional whole.

So, what's a stylesheet?

- **Stylesheet**: A set of rules for consistently styling the content layer. It makes your content layer independent of your presentation layer, so that one can be updated without affecting the other. It governs the visual design of each element within a document, rather than specifying it for each element. For example, in a stylesheet, you can have a rule that says, "All heading 1s must be 18-point Arial bold," instead of having to manually style each individual heading 1 across multiple documents.

14.3 [Theory] Stylesheets

Think of stylesheets as an additional layer overlaid on the layers of content, structure, and presentation.

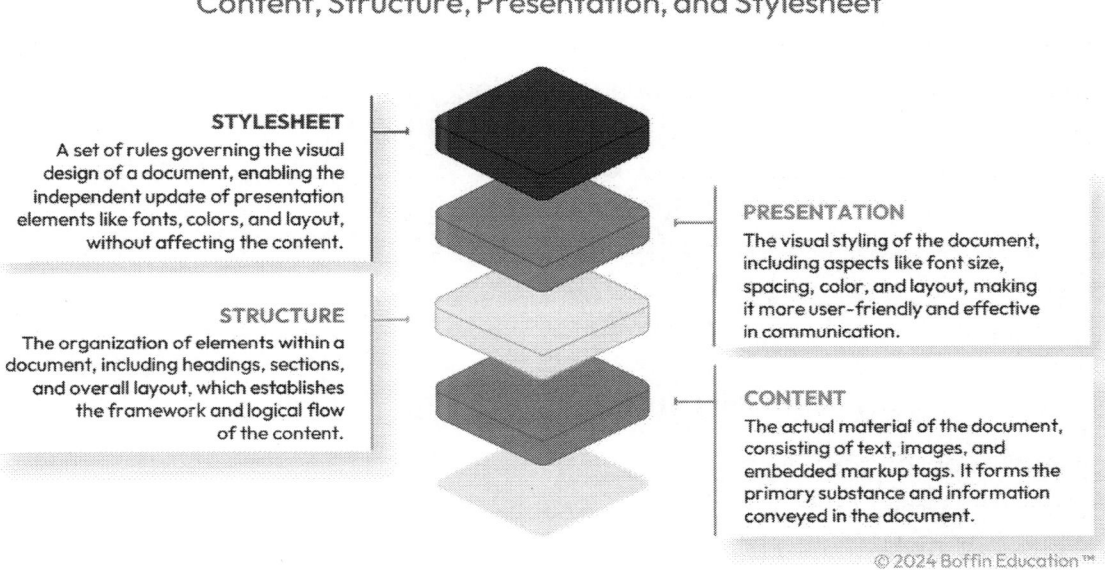

Figure 32: Design Layers of a Document—Content, Structure, Presentation, and Stylesheet

Here are some examples of the many benefits of stylesheets:

- They enable writers to apply visual elements such as fonts, layouts, and sizes across numerous documents without altering the underlying content.
- They simplify any changes from a slight design adjustment, avoiding it having to be a fully-fledged redesign of multiple documents.
- They facilitate the application of a consistent visual brand across numerous documents.
- They facilitate the repurposing of content into multiple formats, like from a content management system into Adobe PDF, Microsoft Word, HTML, and so on.

Chapter 14 Design Stylesheet

 What Does That Mean?

Stylesheet

A set of rules governing the visual design of a document, enabling the independent update of presentation elements like fonts, colors, and layout, without affecting the content.

 Insight

Form vs. Function: An Age-Old Conundrum

The separation of content and presentation taps into an age-old debate: form vs. function. The Greek philosopher Aristotle separated verbal communication into *logos* (the logic of the argument) and *lexis* (the style and delivery of the speech). This dichotomy is mirrored in technical writing through the separation of content (function) and presentation (form). While some argue that the fusion of style and content is an inseparable art form,[32] for technical writers, maintaining a clear distinction is essential. The separation allows for efficient updating of documents, ensuring that changes in design (form) don't hinder the management of the information (function).

14.4 [Theory] Contrast, Alignment, Repetition, Proximity (CARP)

Having a good presentation layer doesn't just make your documentation look better—it helps users grasp the information more effectively by drawing on principles rooted in research.

Although the visual design process might seem to an outsider like an unruly exercise in unbridled creativity, most graphic designers are taught to adhere to design principles, which guide their work. We like the design principles in Robin Williams' popular book, *The Non-Designer's Design Book*,[33] for their memorability and ease of use.

Williams' design principles can be memorized using the acronym CARP: Contrast, Alignment, Repetition, and Proximity. They apply just as much to the design of the presentation layer of a technical document as they do to the design of a house—or pretty much anything else. Once you understand these fundamental principles, you'll start to notice them everywhere.

We've used a worked example of our lead writer, Swapnil's, business card to illustrate the concept. We overlay each of the CARP principles in turn to transform it from an unformatted document into a visually appealing, user-friendly piece of content that successfully achieves its purpose: facilitating information-sharing with a professional contact.

14.4.1 Starting Point: Unformatted

Let's start with an unformatted business card. Think of this as our base content layer, onto which we'll overlay a presentation layer using the CARP principles. The unformatted version has all the correct components that a business card should have, including a logo, but it's almost entirely devoid of design. As a consequence, it's not very usable. It's hard to visually distinguish the different items on the page:

Figure 33: Swapnil's Unformatted Business Card

14.4.2 C: Contrast

What Is Contrast? Contrast makes a visual design element distinguishable from other objects and prevents it from blending into the background. Contrast directs the reader's attention and makes certain elements stand out. This facilitates quick navigation and helps the audience speedily identify the most important information. You can achieve contrast by varying color, size, shape, and typography to highlight key parts of a document, such as titles and headings, making them easily noticeable.

Worked Example: In our worked example, we've applied contrast to make certain elements of Swapnil's business card more prominent:

- **Increased Font Sizes**: The font size for Swapnil's name and job title is increased to make these elements immediately noticeable. These are the most important details that Swapnil thinks should stand out the most, along with his logo.
- **Different Typeface**: The typeface for Swapnil's name is different from the rest of the business card, making it stand out even more.

14.4 [Theory] Contrast, Alignment, Repetition, Proximity (CARP)

- **Use of Color**: Swapnil's job title is highlighted in orange, immediately differentiating it from his name and the other elements on the page. This indicates that it's a vital detail for Swapnil that he wants to direct the reader's attention to.

- **Use of Capitalization**: The headings—e.g., job title, email address, and so on—have been capitalized, differentiating them from the information they contain. This helps make the card more scannable so a reader can quickly find the information they're after, such as a website address.

- **Use of Bold**: Additionally, the subheadings have been highlighted using bold. This makes them even more easily distinguishable from the information they contain.

Notice how this simple technique immediately makes the key information much easier to distinguish? Using contrasting font types, colors, and emphases (bold, italics, and so on) provides a clear visual cue to the document's structure. The headings and subheadings form the information hierarchy, even in a straightforward document like a business card.

Figure 34: Swapnil's Business Card with Contrast

Chapter 14 Design Stylesheet

 Note

Note on Color Usage

When applying color for contrast, it's important to consider how your design will be perceived under different circumstances. Not everyone perceives colors in the same way, and your document may be printed in black and white, potentially diminishing the intended contrast. For example, using red for certain elements and green for others might be effective in a color design, but if these are printed in grayscale, they might appear similar, losing their distinctive meanings. Always ensure that your use of contrast remains effective and clear under various viewing conditions.

14.4.3 A: Alignment

What Is Alignment? Alignment is the practice of lining up design elements along an invisible line—left, right, or center. Strong alignment doesn't just create a striking, neat-looking effect; it complements the way we scan and read pages to find the information we're after. In particular, using left alignment complements the way we naturally visually process information, making for more user-friendly documents. This is called the F pattern, and it's explained in the note below.

Worked Example: In our worked example, strong alignment has been applied to improve readability and make it appear less cluttered:

- **Left Alignment**: The two distinct groupings of information on Swapnil's card have been left-aligned. This creates a neat, two-column effect, distinguishing Swapnil's name and logo from his contact details.
- **Top Alignment**: The left and right columns have been aligned to the same invisible top line. This makes the columns more visually connected, so they look like they're related components of the same document.

14.4 [Theory] Contrast, Alignment, Repetition, Proximity (CARP)

Figure 35: Swapnil's Business Card with Contrast and Alignment

 **Insight**

Alignment and the Visual Patterns of Reading

The alignment principle in CARP is supported by research on the visual patterns of reading. For more detailed insights, see Chapter 14: Design Stylesheet: Visual Patterns of Reading below.

14.4.4 R: Repetition

What Is Repetition? Repetition is the reuse of a visual element consistently, such as color, shapes, typography, and other elements, to establish a theme. It establishes visual coherence and alignment with brand identity. In corporate documents and templates, the organization's colors are often echoed in headings and backgrounds to build a unified visual impression. Typically, a color palette derived from the logo is reused across corporate templates and marketing materials to strengthen branding. This palette is usually found in brand guidelines or as preset styles in templates.

Worked Example: In our worked example, repetition has been applied to reinforce Swapnil's branding:

- **Use of Color**: The orange and black colors of Swapnil's business card create a striking visual design. The same colors have been used to create a color bar running across the top and bottom of Swapnil's business card, reinforcing his logo.
- **Use of Shapes**: The black bars in Swapnil's logo are a design element that has been reused in the color bars on the top and bottom of his card. This creates a strong visual connection with his logo and helps establish a brand.
- **Use of a Watermark**: Swapnil's logo has been repeated as a watermark, fading into the bottom of the card. This creates another element that visually reinforces his logo and brand.

Figure 36: Swapnil's Business Card with Contrast, Alignment, and Repetition

14.4.5 P: Proximity

What Is Proximity? Proximity is the close physical grouping of related elements to establish a visual relationship between them. In essence, proximity implies a relationship in the mind of the audience. Conversely, if you don't want things to appear related, use white space to keep them visually distant.

14.4 [Theory] Contrast, Alignment, Repetition, Proximity (CARP)

Worked Example: In our worked example, we've used proximity and separation to group related items and differentiate unrelated ones:

- **Separation of Name and Logo from Contact Details**: The card has now been split into two pages, one for Swapnil's name, title, and logo, and the other for his contact details. This visual separation distances Swapnil's branding from the sundry details of how to contact him, allowing his brand to make a more dramatic visual impact on its own.

- **Grouping of Items**: On the second page, you'll notice that the contact details have been grouped into the left column, while his call to action has been grouped on the right. This establishes two separate zones—one for contact details and one for his call to action. Additionally, the call to action uses the Z pattern of reading, being placed in the bottom right-hand corner of the document.

- **Use of White Space**: Within each zone, white space has been used to differentiate each subsection from the others, separating things like physical mailing addresses from email addresses. The use of white space creates subzones for distinct subcategories of information.

Chapter 14 Design Stylesheet

Figure 37: Swapnil's Business Card with Contrast, Alignment, Repetition, and Proximity

14.5 [Theory] Visual Patterns of Reading

In *The Design Manual*, David Whitbread explains that research shows how we tend to read on-screen in geometric patterns.[34] This reinforces the importance of strong alignment—the A in CARP—it's not just visually appealing; it complements how we visually process a page.

14.5 [Theory] Visual Patterns of Reading

- **The F Pattern**: Western audiences tend to read in a left-to-right, top-down F pattern: we scan the top of a page, repeat this again a little lower, and then scan downwards vertically along the left side of the page.
- **The Layer-Cake Pattern**: A variation of the F pattern where we scan the headings on a page, in an F pattern, and ignore the text until we find what we're looking for.
- **The Z Pattern**: The way we tend to scan image-heavy pages, using a Z pattern to quickly scan the relevant information. Marketing professionals understand this effect, which is why they often place call-to-action buttons (such as "Buy Now") on the bottom right.

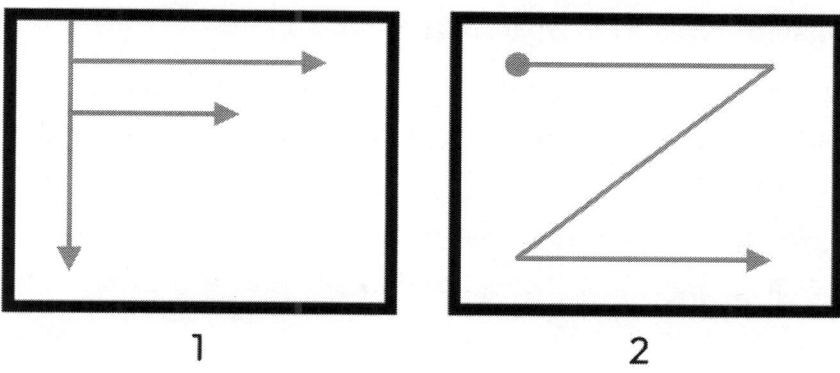

Figure 38: *The F Pattern (1) and the Z Pattern (2) of On-screen Reading (Whitbread, D., 2023)*

Which Pattern Works Best?

The Z pattern is most effective for stand-alone cards, ads, or landing pages—layouts viewed in their entirety without scrolling. The F and Layer-Cake patterns are better suited for pages that extend downward with scrolling.

14.6 [Practice] Design Stylesheet

We've previously discussed the importance of separating presentation from content. This involves using a stylesheet to apply formatting across the document, rather than formatting each element individually. A stylesheet contains formatting rules for predefined document elements such as headings, body text, tables, and lists.

No matter the tools used, the underlying principle remains the same: a stylesheet, often stored as a separate file from the content, applies presentation. This allows for independent updates. Creating a stylesheet is an advanced technical writing skill. It requires interpreting and applying your organization's brand guidelines in collaboration with the marketing team. In the absence of brand guidelines, creativity is key. You'll need to draw from elements of your organization's logo, branding, and existing templates to craft a new stylesheet.

The steps in designing a stylesheet are as follows, with support provided by our template Visual Design Principles Checklist. The next section will include a worked example of these steps:

1. **Identify Key Elements to Style**: List and describe common document elements such as headings, body text, tables, and lists that will need consistent styling.

2. **Choose Your Tools**: Select the appropriate tools, like Microsoft Word or CSS, based on project requirements or your organization's preferred technologies.

3. **Apply Brand Guidelines**: Analyze the brand guidelines to understand your organization's presentation requirements. Collaborate with marketing teams as needed. If no guidelines are available, use your organization's logo, branding, and templates as inspiration.

4. **Develop Styles for Each Element**: Craft styles for each key element, considering font size, color, alignment, and so on. Keep the CARP principles (Contrast, Alignment, Repetition, Proximity) in mind as a design guide. Refer to our template for detailed considerations: *Visual Design Principles Checklist*.

5. **Test and Iterate**: Test the stylesheet across formats and devices, ensuring it works for all publication formats.

6. **Publishing**: Release the stylesheet into your team or organization's template library for use.

 Insight

Did Mobile Kill My Docs? Cross-Platform Design

When developing technical documentation, it's important to consider not just the content, but also the platform through which your audience will engage with it. The choice between a desktop and a mobile device brings unique challenges and opportunities that can significantly impact the effectiveness of your documentation.

Here are some key considerations:

- **Screen Size**: Test your documentation on various devices with different screen sizes to ensure accessibility and a positive user experience. The contrast in screen sizes between desktop and mobile devices is significant.
- **Touch vs. Click**: Tailor your documentation for the interaction methods of your audience—mouse clicks for desktop users and touch gestures for mobile users.
- **Tables**: Be cautious with using formats like tables, which may not render well on mobile devices.

By being attentive to how your audience will interact with your content, you can create documentation that is not only informative but also accessible and engaging across the key platforms: desktop and mobile.

14.7 [Sample] Sample Business Card Stylesheet

In the following section, we provide a worked example to illustrate the stylesheet behind Swapnil's business card in Microsoft Word and Cascading Style Sheets (CSS). CSS is a computer language used to lay out web pages in HTML.

Chapter 14 Design Stylesheet

Here are the distinct elements of Swapnil's business card for which we are going to create styles:

1. **Element 1**: Page
2. **Element 2**: Title
3. **Element 3**: Heading 1
4. **Element 4**: Heading 2
5. **Element 5**: QR Code
6. **Element 6**: Emphasis
7. **Element 7**: Body Text (not shown in diagram)

14.7 [Sample] Sample Business Card Stylesheet

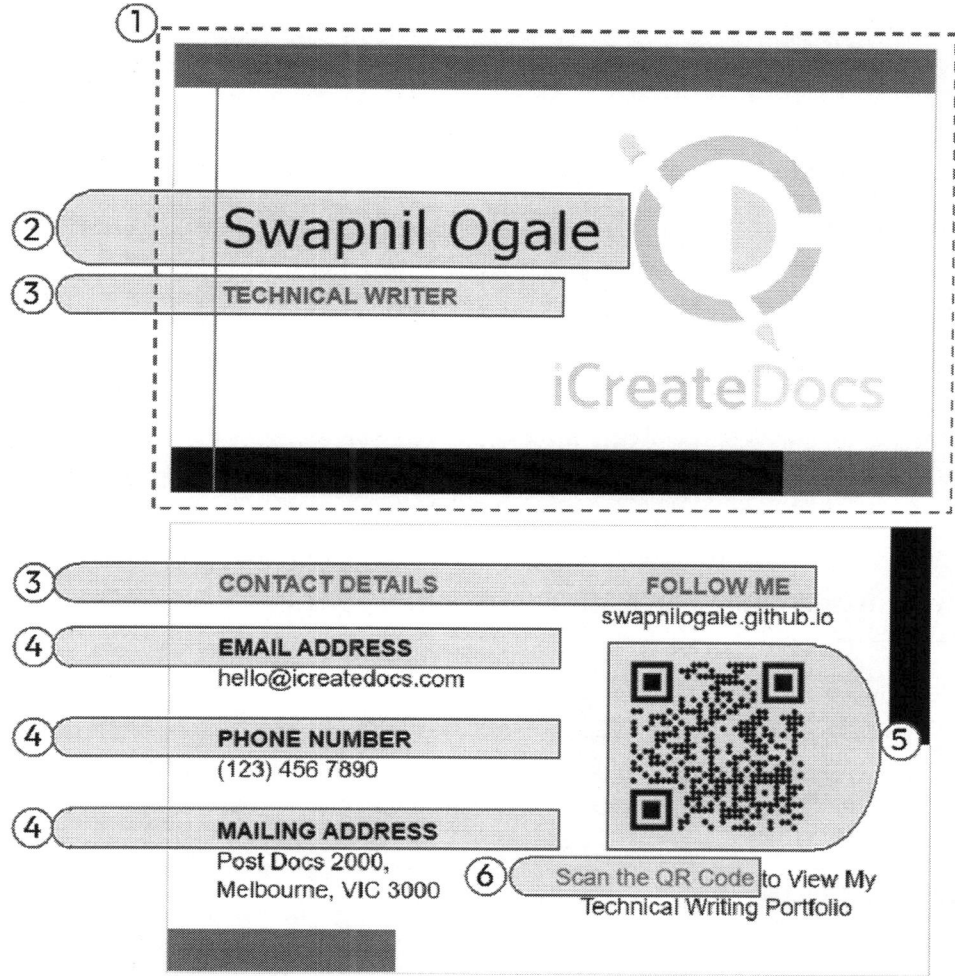

Figure 39: Design Elements in Swapnil's Business Card

Chapter 14 Design Stylesheet

 Note

Using Stylesheets in Different Technologies

- **Microsoft Word**: In Microsoft Word, use the Styles pane to design styles within a document. Save the stylesheet you have created as a Style Set, which you can apply to other documents or use as a template.

- **CSS Stylesheets**: To create a stylesheet in CSS (Cascading Style Sheets), you will need to define styles in a .css file, which can then be linked to your HTML document.

Stylesheet in Microsoft Word	Stylesheet in CSS

Page
 	
- **Margins**: Left: 0.6cm; Right: 0.6cm; Top: 0.6cm; Bottom: 0.6cm
- **Paper**: Width: 9.1cm; Height: 5.5cm
- **Header and Footer**: Header distance from edge: 0cm; Footer distance from edge: 0.5cm
	@media print {
	@page {
	size: 9.1cm 5.5cm;
	margin: 0.6cm;
	
	/* Note: CSS does not support specifying distances for headers and footers from the edge. This is typically controlled by the browser's print settings or the properties of the printing device. */
	
	}
	}

248 | Technical Writing Process 2nd Edition

14.7 [Sample] Sample Business Card Stylesheet

Stylesheet in Microsoft Word	Stylesheet in CSS

Title
 	
- **Paragraph Style**: Title
- **Font**: Verdana, 18pt
- **Alignment**: Left
- **Indentation**: Left: 0cm; Right: 0cm
- **Spacing**: Before: 36pt; After: 6pt; Line spacing: Multiple 1.08
	```.title {```
	`font-family: Verdana;`
	`font-size: 18pt;`
	`text-align: left;`
	`margin-left: 0;`
	`margin-right: 0;`
	`margin-top: 36pt;`
	`margin-bottom: 6pt;`
	`line-height: 1.08;`
	`}`

Heading 1
 	
- **Paragraph Style**: Heading 1
- **Font**: Arial, 8pt, Bold, Custom Color (RGB(209,84,37))
- **Effects**: All caps
- **Alignment**: Left
- **Outline Level**: Level 1
- **Indentation**: Left: 0cm; Right: 0cm
- **Spacing**: Before: 0pt; After: 0pt; Line spacing: Multiple 1.08
	`.h1 {`
	`font-family: Arial;`
	`font-size: 8pt;`
	`font-weight: bold;`
	`color: rgb(209, 84, 37);`
	`text-transform: uppercase;`
	`text-align: left;`
	`margin-left: 0;`
	`margin-right: 0;`
	`margin-top: 0;`
	`margin-bottom: 0;`
	`line-height: 1.08;`
	`}`

Stylesheet in Microsoft Word	Stylesheet in CSS
| Heading 2<br><br>- **Paragraph Style**: Heading 2<br>- **Font**: Arial, 8pt, Bold<br>- **Effects**: All caps<br>- **Alignment**: Left<br>- **Outline Level**: Level 2<br>- **Indentation**: Left: 0cm; Right: 0cm<br>- **Spacing**: Before: 12pt; After: 0pt; Line spacing: Multiple 1.08 | ```
.h2 {
font-family:       Arial;
font-size:           8pt;
font-weight:        bold;
text-transform: uppercase;
text-align:          left;
margin-left:         0;
margin-right:        0;
margin-top:       12pt;
margin-bottom:       0;
line-height:       1.08;
}
``` |
| Emphasis<br><br>- **Character Style**: Emphasis<br>- **Font**: Arial, 8pt, Custom Color (RGB(209,84,37)) | ```
.emphasis {
font-family:       Arial;
font-size:           8pt;
color:   rgb(209, 84, 37);
}
``` |
| QR Code<br><br>- **Paragraph Style**: QR Code<br>- **Font**: Arial, 8pt<br>- **Alignment**: Centered<br>- **Indentation**: Left: 0cm; Right: 0cm<br>- **Spacing**: Before: 10pt; After: 10pt; Line spacing: Multiple 1.08 | ```
.qr-code {
font-family:       Arial;
font-size:           8pt;
text-align:        center;
margin-left:         0;
margin-right:        0;
margin-top:       10pt;
margin-bottom:    10pt;
line-height:       1.08;
}
``` |
```

14.7 [Sample] Sample Business Card Stylesheet

Stylesheet in Microsoft Word	Stylesheet in CSS

Body Text (Normal)
 	
- **Paragraph Style**: Normal
- **Font**: Arial, 8pt
- **Indentation**: Left: 0cm; Right: 0cm
- **Spacing**: Before: 0pt; After: 0pt; Line spacing: Multiple 1.08
	.body-text {
	font-family: Arial;
	font-size: 8pt;
	margin-left: 0;
	margin-right: 0;
	margin-top: 0;
	margin-bottom: 0;
	line-height: 1.08;
	}

Chapter 15 Design Templates

Lead Writers: John New, Kieran Morgan | **Expert Reviewer/s**: Deirdre Wilson

This chapter focuses on the role of templates in technical documentation, distinguishing them as unique technical documents through the use of instructional, placeholder, and sample text. It presents a practical, step-by-step approach to template creation with a focus on maintaining consistency and aligning with organizational branding. The chapter also highlights the role of generative AI in streamlining the template creation process and includes both a sample user guide template and a comprehensive checklist to facilitate effective template design.

 Who Should Read This

- Beginner Technical Writers
- Career Advancers
- Cross-Domain Professionals

CONTENTS

15.1 Introduction	255
15.2 [Theory] Forms vs. Templates	256
15.3 [Theory] Metainformation, the Unique Characteristic of Templates	257
15.4 [Theory] Common Elements of Templates	258
15.5 [Practice] Design Templates	259
15.6 [Template] Technical Document Template	261

Chapter 15 Design Templates

PROCESS

Inputs	Design > Design Templates		Outputs
• Brand Guidelines • Best-Practice Examples	**Theory**	• Forms vs. Templates • Metainformation, the Unique Characteristic of Templates • Common Elements of Templates	• Document Template
	Practice	• Plan • Design • Write • Review • Publish	
	Examples	• N/A	
	Templates	• Technical Document Template • Template Design Checklist	

15.1 Introduction

Imagine you're baking a cookie or a cake. You reach for a cookie cutter to shape your dough before baking. This simple tool provides a predefined shape that makes your task easier and the outcome more consistent. Similarly, in technical writing, templates serve as indispensable tools, just like cookie cutters. They offer an established mold or pattern for your documents.[35]

Templates are essential tools for all technical writers. They save time by eliminating the need to design and create a new document from scratch. They also ensure alignment with your organization's branding and maintain consistency across writers. A well-designed template allows you to focus on content development rather than on formatting and design, speeding up the writing process. Additionally, templates guide writers toward uniformity by offering a consistent structure with predefined elements and placeholder text, which helps to facilitate a standardized layout and language across the organization.

This chapter guides you through the process of creating a template from scratch. It explains how templates are structured into three main elements: instructional text, placeholder text, and sample text. These unique elements constitute the "metainformation" layer, which distinguishes templates from all other document types. We will then discuss the distinction between forms and templates, and explain some of the common elements expected in templates.

Finally, we offer practical steps to guide you through the process of template creation, together with a handy Template Design Checklist.

> **Insight**
>
> **Check Before You Create**
>
> For technical writers in mature organizations, starting a document without a template is rare. If you've been asked to write a document and no suitable template exists, make sure you double-check before creating one from scratch. If there really is none, it's a prudent step to create one in case a similar document type is needed in the future.

15.2 [Theory] Forms vs. Templates

When you're designing templates, it's important to understand the distinction between forms and templates, as they serve different purposes:

- **Forms (Information Gathering)**: Forms are designed to collect data or information. They are often interactive and designed to be user friendly, guiding the user to provide specific types of information. Forms can be physical or digital, like interactive webpages or even Adobe PDFs, and are important in situations where structured data collection is essential. They are particularly useful in gathering data and ensuring that the information collected is organized and consistent.

- **Templates (Cookie Cutters for Information)**: Templates are pre-structured documents used to maintain consistency and efficiency in document creation. They serve as a framework for presenting information, ensuring that all documents adhere to a specific format, style, and structure. Templates often include instructional text, placeholders, and sample text to guide the user in creating the final document. They are essential for branding, ensuring consistency across documents, and saving time and effort in document creation.

> **What Does That Mean?**
>
> **Forms**
>
> Interactive tools designed for structured data collection, guiding users in providing specific information. Forms can be physical or digital, and are important for organizing data input in an efficient and consistent manner.
>
> **Templates**
>
> Predesigned frameworks for documents, providing standardized layouts and styles to ensure consistency and efficiency in writing while aligning with an organization's branding and visual identity.

15.3 [Theory] Metainformation, the Unique Characteristic of Templates

Templates are just like any other technical document—with one very important difference. Templates have a unique set of characteristics that turn them from cookies into cookie cutters. This is what we call the metainformation layer. It's what makes templates a class of document all on their own.

The metainformation layer consists of three elements:

- **Instructions**: Instructions guide users in the proper and consistent use of the template. They are there as a temporary signpost to guide the template creation.
  - **How to Identify**: Instructions are typically placed on a clearly marked cover sheet or textbox at the front of a template so they can be deleted prior to publication.
- **Placeholder Text**: Placeholder text acts as a prewritten script for content within the document. It guides users toward consistency of language, where a premium is placed on consistency rather than creativity.

- **How to Identify**: Placeholder text is usually differentiated from other text by the use of [square brackets] for variables, a cue for the user to replace them with their own text.

- **Sample Text**: Sample text goes a step beyond placeholder text to provide an additional layer of guidance. Sample text gives the template user an idealized example of how a section could be worded, often drawing from previous examples considered best practice.

  - **How to Identify**: Sample text is often identified by italics or a different font color.

 What Does That Mean?

Metainformation

Metainformation is information about other information. This is a term we've used to describe instructions and placeholder text in templates. In doing so, we've leaned on the definition of a closely related concept, metadata, which is "data that provides information about other data."[36] If you're not clear on the difference between data and information, see the DIKW Pyramid in Chapter 9: Collect Information.

15.4 [Theory] Common Elements of Templates

In addition to metainformation, a well-designed template includes various elements that contribute to its effectiveness and ease of use. These elements ensure comprehensiveness, consistency in the documents created from the template, and adherence to best practices.

- **Style Sheets**: Defined styles for elements such as headings and paragraphs are essential for maintaining consistency in formatting and appearance, as well as for alignment with branding elements like logos and color schemes. See Chapter 14: Design Stylesheet.

- **Document Control Tables**: These are important for maintaining version control and document history. Typically these tables include information on document revisions, authorship, approval dates, and change history.

- **Predefined Headings**: Consistent sections in the templates provide a structured approach that end users expect in the documents.

- **Footer and Header Information**: Standardized headers and footers, containing the document title, page numbers, copyright statement, version number, and other relevant information, are important for professional presentation and navigation.

- **Tables and Charts**: Preformatted tables and charts facilitate uniform data representation across various documents.

- **Legal and Compliance Notes**: These sections accommodate disclaimers, copyright information, and other legal requirements, ensuring compliance and protection.

15.5 [Practice] Design Templates

Designing templates is a senior technical writing skill. It involves skillfully orchestrating technical writing theory and practice in concert. In fact, designing a template is a microcosm of the entire technical writing process. The only thing that templates don't normally require is a comprehensive documentation plan—unless they'll be used by hundreds of users, in which case, perhaps it's best to consider one.

Here are the steps to create a template, organized by the phases of the technical writing process.

15.5.1 Step 1: Plan

- **Audience Analysis**: Determine who will use your templates, and consider their skill level and needs. This will guide the template's usability.
- **Obtain Guidelines**: Obtain your organization's brand guidelines or other rules concerning the visual layout of documents.
- **Collect Examples**: Gather best-practice document examples to inform your template design. This can include internal documents, such as existing templates, as well as external sources for broader insights.

15.5.2 Step 2: Design

- **Structure Development**: Create a hierarchy of headings based on common elements found in your best-practice documents. This sets the foundational structure of your template.
- **Stylesheet Creation**: Develop a stylesheet following your organization's brand guidelines or adapt an existing organizational stylesheet.

15.5.3 Step 3: Write

- **Placeholder Text**: Write placeholder text under each heading, balancing specificity and generality. Use clear markers like [square brackets] for identification.
- **Sample Text**: Compose sample text that provides practical guidance and examples that are distinctively marked for easy recognition.
- **Instructional Text**: Explain template usage at the front, covering placeholder text identification and deletion. Consider linking to an intranet page for detailed instructions.

15.5.4 Step 4: Review

- **Self Review**: Evaluate the template using the Template Design Checklist for thoroughness.
- **Testing**: Test the template on a real use case and seek peer feedback.
- **External Review**: Have your brand and legal teams review the template for compliance and accuracy.

15.5.5 Step 5: Publish

- **Publishing**: Secure approval and publish your template. Aim for wide communication about its use, especially if it supersedes an older version. Provide clear guidelines on its usage and whom to contact for inquiries.

 Tip

Use Artificial Intelligence to Streamline Template Creation

If your organization is okay with you using generative AI, it can save you a lot of time in designing templates. One of the greatest strengths of generative AI is its ability to synthesize existing information, design structure, and write placeholder text—all of which are extremely helpful for template creation. See Chapter 8: Artificial Intelligence (AI) for Technical Writers for more information on how to use AI tools such as ChatGPT.

15.6 [Template] Technical Document Template

To demonstrate the application of the principles in this chapter, we've created a Technical Document Template. It's a general-purpose template that you can use to create any technical document in Microsoft Word. This template features a minimalist stylesheet specifically tailored for technical documents.

It incorporates the metainformation elements described above in the following ways:

- **Instructional Text**: Clearly marked instructions are situated at the beginning of the template and within relevant headings. These provide guidance for users on how to use the template and should be deleted prior to publication, as per the guidance in the template.

- **Placeholder Text**: Placeholder text, marked with [square brackets], indicates to the user where they should insert their own variables into prewritten sentences. This guides users toward uniformity in places where consistency is valued over creativity.

- **Sample Text**: Our template provides essential placeholder text, such as sample headings in the introduction, as well as standard elements like a document control table and copyright statements. This gives users a concrete example of what's expected in each section without being too prescriptive.

Part 6 Write

Mastering the art of writing in technical documentation.

Chapter 16 Writing Principles

Lead Writer: Steve Moss | **Peer Reviewer/s**: Kieran Morgan

This chapter explains essential principles of the writing aspect of technical writing. It covers strategies for empathizing with readers, organizing complex information, and crafting polished prose. Each section offers practical examples, dos and don'ts, and advice on applying these principles within the framework of your style guide and house style manual. This is a comprehensive guide for technical writers seeking to elevate the quality and efficiency of their work.

 Who Should Read This

- Aspiring Technical Writers
- Beginner Technical Writers
- Cross-Domain Professionals

CONTENTS

16.1 Introduction 267
16.2 [Theory] Empathizing with Your Audience 268
16.3 [Theory] Organizing Information 271
16.4 [Theory] Perfecting Prose 278

Chapter 16 Writing Principles

PROCESS

Inputs	Write > Writing Principles		Outputs
• Style Guide • House Style Manual	**Theory**	• Relevance • KISS—Keep It Simple • Brevity • Chunking • Signposting • Layering • Bulleting • Verb–Noun Structure • Active Voice • Imperative Mood • Parallel Language • Write to the Reader	• Well-Written Drafts

16.1 Introduction

Step into the world of masterful technical writing with this chapter. Drawing on the wisdom of technical writers with decades of experience, we focus on the core principles underpinning high-quality, efficient technical writing. You'll learn how to communicate complex information effectively to your audience. This chapter provides insights on applying these principles, enriched with practical examples as well as the dos and don'ts of the craft.

Your guides on this journey are your style guide and house style manual. These indispensable tools aid in perfecting the pesky finer points of technical communication, such as the proper placement of commas in a bulleted list.

 What Does That Mean?

Style Guide

A manual of guidelines on grammar, punctuation, layout, formatting, structure, and other stylistic aspects of writing. An example is the *Chicago Manual of Style* in the United States. Style guides vary from country to country and across different industries.

House Style Manual

An organization-specific (or even team-specific) style guide. House style manuals contain guidelines on how to write in the organization's style, dos and don'ts, trademarks, legal boilerplate, the correct tone to adopt, and other relevant details.

16.2 [Theory] Empathizing with Your Audience

In this section, we explain audience-centric technical writing. The key to crafting excellent technical documentation lies in empathizing with your audience. This segment builds upon the foundation laid in our discussion on audience analysis in Chapter 11: Analyze Audience.

> **Insight**
>
> **The Technical Writer as an Audience Advocate**
>
> In technical writing, think of yourself as the audience's advocate. Your job isn't just to relay information; it's to ensure the information is understandable and relevant from the reader's point of view. This means constantly asking, "Will this make sense to end users?" By adopting this mindset, you bridge the gap between expert knowledge and the reader's needs to make complex information accessible and useful.

16.2.1 Key Principle: Relevance

Imagine you're a curator tasked with assembling only the most compelling pieces for an exhibition. That's your role when applying the relevance principle in technical writing. Your goal is to sift through information and select only what is most relevant for your audience. This principle isn't just about what to include but also what to leave out. Because you've gathered a wealth of information from subject matter experts and existing technical documents, it's tempting to showcase it all. Resist the temptation! Great writers use only what's absolutely necessary.

Consider the following when applying the relevance principle:

1. **Relevance to the Section**: Does the information enhance the specific topic you're addressing at this moment?

2. **Overall Document Relevance**: Does it fit the document's purpose, offering value to your audience regarding the product or service in question?

When to Use:

- While reviewing materials gathered during the information collection stage
- When evaluating the contents of a document, either in development or already existing
- Deciding on the placement of information within a document's structure

Table 19: Relevance Dos and Don'ts

Do	Don't
Place valuable information where it most logically fits or note it for later inclusion.	Include information without assessing its utility in context.
Consult with SMEs to validate the relevance and placement of information if you're unsure.	Hesitate to exclude information that doesn't serve the document's purpose.

 Insight

Developing Your Relevance Radar

Develop your own "relevance radar" by continuously asking yourself whether the information is necessary for the reader at the current point in the document or even in the document as a whole. If you're in doubt about a piece of information's relevance, provisionally include it and seek subject matter expert validation during the next review. Their feedback is crucial in discerning the value of contentious information.

16.2.2 Key Principle: KISS—Keep It Simple

The KISS principle, an acronym for "Keep It Simple, Stupid!," emphasizes the value of simplicity in communication. Originally used in the military to ensure straightforward procedures, this principle is equally important in technical writing. It advocates for writing in a manner that is easily understandable to the average reader, without assuming they have prior knowledge or training. The goal is to make complex information accessible and usable, rather than underestimating your audience's intelligence. Even experts begin as beginners! The real challenge lies in making your content clear and straightforward so it serves a broad audience.

When to Use:

- In situations where the audience might lack background knowledge
- To avoid overloading the reader with unnecessary jargon or details

Table 20: Assumptions Not to Make about Your Audience

Assumption to Avoid	Reason
They have received training in the product, process, or procedure.	Many may be new to the topic.
They have been with the company for some time.	New employees might be your audience.
They understand company jargon.	Jargon can be confusing and exclusive.
They want extensive documentation.	Most prefer concise, to-the-point information.
They are familiar with the document's history and context.	Background knowledge can't be assumed.

Assumption to Avoid	Reason
They know where to find supporting tools and documents.	Accessibility of resources varies.

16.3 [Theory] Organizing Information

In this section, we explore how to structure your writing to enhance clarity and comprehension. We'll reveal strategies for organizing complex information to ensure your audience can easily navigate and understand your content. By understanding these principles, you'll learn to create a logical flow that guides your readers through complex technical details with ease.

16.3.1 Key Principle: Brevity

"I apologize for such a long letter—I didn't have time to write a short one."

—Mark Twain

Writing clear, concise content takes time and effort, but your readers will thank you. It's a balancing act: being concise without sacrificing meaning. The principle of brevity isn't just about using fewer words; it's about making each word count. The irony is that writing succinctly is often more challenging than filling pages with words. Crafting concise prose is an art that requires writers to distill complex information into easily digestible content. Remember, great writers often spend more time cutting down their work than building it up, ensuring every detail serves a purpose and nothing more.

When to Use:

- In all levels of document creation, from sentences to full sections
- Especially useful in documents that will be translated, as shorter, clearer sentences reduce translation complexity

Chapter 16 Writing Principles

Table 21: Brevity Dos and Don'ts

Do	Don't
Start with longer sentences, then edit down.	Assume the first draft will be anything like the final version.
Aim for sentences averaging 15–20 words.	Create overly short, disconnected sentences.
Break down complex sentences into shorter ones.	Sacrifice clarity for the sake of brevity.

 Insight

Reading Effort Is the Inverse of Writing Effort

Effort in writing inversely correlates with effort in reading. The more you refine your writing, the easier it is for your audience to understand. This statement reflects the technical writer's craft, where they make the effort to capture the essence of the material and present the reader with only what is needed. Unfortunately, some people imagine that if a text is easy to read, it was also easy to write, and therefore anyone could do it. This is certainly not the case.

16.3.2 Key Principle: Chunking

Chunking is an essential technique in technical writing that makes information more accessible and memorable for readers. It involves organizing content into manageable units or "chunks." This concept is rooted in cognitive psychology, drawing particularly from the work of George A. Miller[37] and, later, Nelson Cowan[38] who studied the capacity of human short-term memory. Although the exact number of items that constitute an optimal "chunk" varies—Miller suggested seven plus or minus two, while Cowan proposed about four—the principle remains consistent: reduce cognitive load by presenting information in small, digestible segments. This approach is particularly effective when you're structuring chapters, bulleted lists, or procedural instructions.

When to Use:

- When organizing topics within a chapter to avoid overwhelming the reader
- For creating bulleted or numbered lists, particularly in procedural contexts
- During the initial drafting or subsequent review of document structure to ensure information is presented in an easily digestible format

Table 22: Dos and Don'ts of Chunking

Do	Don't
Aim for an average chunk size of 5±2 items.	Overload sections with too much information.
Group related items under a common heading.	Arbitrarily force splits in content merely to stay within the 5±2 items chunk size.
Adapt chunk sizes based on content needs.	Ignore the natural flow of information and its nuanced complexity.

> **Note**
>
> **Don't Force Splits in Content**
>
> Don't twist your writing into knots trying to force a split just to satisfy the rule of five plus or minus two items. If a bulleted list contains, say, nine items, or a chapter encompasses eight topics, and there's no practical way to split these into smaller chunks, don't worry. Arbitrarily dividing a number of items—by forcing in a heading—can be worse than presenting too many items. There will be plenty of other opportunities to serve your reader by applying the chunking technique!

16.3.3 Key Principle: Signposting

Signposting involves the use of clear and meaningful headings at various levels within a document. This principle is important because technical documents usually serve as reference materials rather than as texts that are read from beginning to end. Effective signposting allows readers to quickly scan a document to locate the information they need. It requires crafting titles that accurately reflect the content and purpose of the document, its chapters or sections, individual topics, and even specific paragraphs or information blocks. These blocks may include text, tables, or visual elements.

When to Use:

- When titling the entire document to reflect its overall purpose
- When naming chapters or sections to indicate their specific content and purpose
- When heading topics within sections to guide the reader to relevant information
- In concert with verb–noun structure for headings in process-based documentation
- When labeling paragraphs or information blocks to summarize the enclosed content, whether it's text, tables, or visual elements

16.3.4 Key Principle: Layering

A helpful side effect of signposting is a technique known as layering. Layering enhances navigability and user engagement by structuring content into distinct, hierarchical sections. This method involves using clear headings or titles at various levels of a document to create layers. Each layer represents a specific topic or level of detail. By doing so, readers can easily skim through the document to find the information they need without reading every detail. This approach is especially useful in lengthy or complex documents where users might seek specific information, like a particular procedure or definition.

When to Use:

- In lengthy documents or manuals to enhance readability and navigation
- When the content includes varied topics or subtopics that can be categorized distinctly
- In online help systems or digital documents to facilitate efficient search and navigation
- When the audience is diverse and their needs for information vary in depth and specificity

---

Example 1: Layering in a Manual

Context: A printed manual on data management.

Layer Structure:

- Chapter: "Data Management"
- Section: "Archiving"
- Topic: "Archiving Client Data"

Usage: Allows readers to quickly locate the section on Archiving and then drill down to the specific topic of Archiving Client Data.

> **Example 2: Layering in Online Help**
>
> **Context**: An online help system for a software application.
>
> **Layer Structure**:
>
> - Searchable Phrase: "Archive client data"
> - Menu Hierarchy: Home > Help > Data Management > Archiving > Client Data
>
> **Usage**: Users can type "archive client data" in the search box or navigate through help menus to find detailed instructions on the topic.

16.3.5 Key Principle: Bulleting

Bulleting is a technique employed to enhance the readability and memorability of lists. Unlike sentences that use commas or semicolons to separate items, bulleting presents each item on a new line. This format is visually more appealing and simplifies information processing for the reader. It is particularly effective in technical writing, where clarity and accessibility are of paramount importance.

When to Use:

- Presenting multiple options or values
- When the list items are distinct and need to be emphasized
- In situations where quick reference or memorability is essential

16.3 [Theory] Organizing Information

Example: Comma-Separated List	Example: Bullet Point List
"Our premium holiday rental property has four bedrooms, two bathrooms, a spacious living area, a carport, and a swimming pool."	"Our premium holiday rental property has: • four bedrooms, • two bathrooms, • a spacious living area, • a carport, and • a swimming pool."

Table 23: Dos and Don'ts of Bullet Points

Do	Don't
Use a meaningful introductory sentence or phrase before the list.	Repeat the same word at the start of each bullet; reword if necessary.
Use consistent capitalization and punctuation for each item.	Use bullet points if the sequence of items is important; use numbered lists instead.
Keep line lengths similar for visual harmony.	Create a bulleted list for a single item; incorporate it into the sentence instead.
Limit the list to 5–7 items for better reader retention; use subheadings for longer lists.	Arbitrarily force splits in lists merely to stay within the 5–7 item size.
Use "and" or "or" in the second-to-last item to clarify whether the list is inclusive or exclusive.	

16.4 [Theory] Perfecting Prose

In this section, we aim to refine your writing style for technical documents. You will learn to craft clear, concise, and engaging prose that resonates with your audience. We'll explore the nuances of language choices, sentence structure, and readability, helping you transform technical jargon into accessible and compelling content.

16.4.1 Key Principle: Verb-Noun Structure

Verb–noun structure is a straightforward yet effective method for crafting headings, particularly in process and procedure documentation. By placing a verb before a noun, this structure clearly indicates the action and its object, making it easier for readers to understand the purpose of the information that follows. It's widely used in process modeling and documentation because of its ability to maintain logical consistency and provide a formulaic template for each activity within a process.

For example, in a process model, each activity box named using this method will describe both an action (verb) and the object of that action (noun), thereby eliminating possible ambiguity. This approach also invites questions: Who performed this action (role)? What is the consequence (output) of that action? These questions are essential for the development of fully fleshed-out processes.

16.4 [Theory] Perfecting Prose

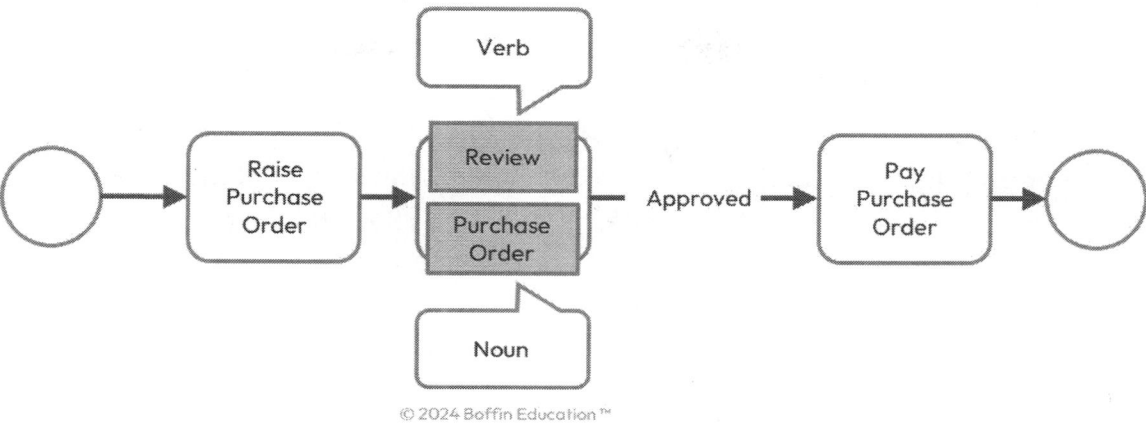

Figure 40: Verb–Noun Structure in a Simple Process

When to Use:

- Creating or reviewing headings for processes and procedures
- Naming activities within a process
- Developing titles for process-related documentation

Table 24: Dos and Don'ts for Verb–Noun Structure

Do	Don't
Use for headings in process documentation.	Apply to descriptions of concepts or theories.
Follow the formula of verb followed by noun.	Overuse in nonprocedural contexts.
Ensure clarity and action-oriented language.	Use for general background information.

Chapter 16 Writing Principles

 Note

Processes vs. Procedures

Processes offer a high-level view of the activities needed to achieve a result (such as processing a purchase order), while procedures detail the specific steps for an activity (such as how to raise a purchase order). Understanding this distinction is key to applying the verb–noun structure appropriately.

 Note

The Technical Writing Process Uses Verb–Noun Structure

You've probably noticed that sections of the *Technical Writing Process* use a verb–noun structure for headings in many chapters. This approach helps us structure the book in tune with its process-oriented nature, making it logical, consistent, and easy for our audience (you) to follow.

16.4.2 Key Principle: Active Voice

Active voice is a writing technique that focuses attention clearly on the "doer" of an action, making your writing more direct and helping to reduce ambiguity. Unlike passive voice, which can obscure the performer of the action, active voice places the subject at the forefront. Consider: "The designer produced an attractive portfolio," as opposed to "An attractive portfolio was produced by the designer." This approach not only clarifies who is doing what but also typically results in shorter, more straightforward sentences. Active voice lends itself naturally to technical writing. It's especially useful in writing processes or procedures, as it minimizes ambiguity and maintains focus on the subject.

16.4 [Theory] Perfecting Prose

Example: Passive Voice	Example: Active Voice
"An attractive portfolio was produced by the designer."	"The designer produced an attractive portfolio."
"Feedback on the project was given by the team leader."	"The team leader gave feedback on the project."

When to Use:

- In most technical writing scenarios, especially when describing actions in processes or procedures
- When clarity about the performer of an action is crucial
- To achieve conciseness in your writing

Table 25: Dos and Don'ts of Active Voice

Do	Don't
Use active voice to highlight the "doer" of an action.	Overuse passive voice, which can obscure the doer.
Use active voice for clarity in process descriptions.	Rely solely on active voice; sometimes passive is appropriate.
Opt for active voice to make sentences more concise.	Forget that in some cases, the doer may be intentionally omitted.

16.4.3 Key Principle: Imperative Mood

The imperative mood in technical writing is about directness, ensuring that instructions are clear. This mood uses a verb form that conveys a command or request, such as "Click the print button." This differs from the subjunctive mood, which suggests possibilities ("I would suggest that you click the print button"), and the indicative mood, which states facts or opinions ("You can click the print button if you like"). The imperative mood is about providing clear, direct actions. It's not about being forceful or rude; rather, it focuses on being helpful and efficient in communication. Technical writing benefits from this direct approach as it guides the reader clearly and reduces ambiguity.

Example: Subjunctive Mood	Example: Indicative Mood	Example: Imperative Mood
"I would suggest that you click the print button."	"You can click the print button if you like."	"Click the print button."
"Consider adjusting the settings for better results."	"You might want to adjust the settings for better results."	"Adjust the settings for better results."

When to Use:

- Writing step-by-step instructions
- Developing user guides or training materials
- Any scenario where clear, concise directions are needed

Table 26: Dos and Don'ts of Imperative Mood

Do	Don't
Place the verb at the start of the sentence.	Use overly polite or indirect language.

Do	Don't
Use the present tense to imply immediate action.	Add unnecessary words that dilute the directness.

16.4.4 Key Principle: Parallel Language

Parallel language, also known as parallelism or parallel structure, is a key principle in technical writing that enhances readability and clarity. It involves using the same grammatical structure throughout a sentence, especially in lists or series. This consistency in structure allows readers to follow the text more easily, providing a rhythm and symmetry that make the content more digestible. Applying parallel language is not just about grammatical correctness; it's about crafting a smooth, cohesive reading experience.

Example: Nonparallel Language	Example: Parallel Language (Gerund Form)	Example: Parallel Language (Infinitive Form)
"My brother likes skydiving, hang-gliding, and to surf."	"My brother likes skydiving, hang-gliding, and surfing."	"My brother likes to skydive, to hang-glide, and to surf."

When to Use:

- When listing items or actions in a sentence.
- In constructing sentences that contain comparisons.
- In bullet points or numbered lists to maintain a uniform structure.
- For headings and subheadings to ensure stylistic consistency.

Table 27: Dos and Don'ts of Parallel Language

Do	Don't
Match the grammatical forms of listed items (all gerunds or all infinitives).	Mix different forms (gerunds with infinitives) in a list.
Apply parallelism in series of actions or descriptions.	Use inconsistent grammatical structures, leading to awkward or confusing sentences.
Use parallel structures in headings and subheadings.	Overlook the need for parallelism in structural elements of your writing.

16.4.5 Key Principle: Write to the Reader

Writing directly to the reader, known as the second-person narrative style, creates a friendly and engaging tone. This approach makes the reader feel like an active participant in the text. Unlike the first person ("I, me") or third person ("he, she, they") perspectives, which tend to be more descriptive and detached, the second person ("you, your") establishes a direct connection with the reader. It's particularly effective in technical writing, where clarity and direct instruction are paramount. The second-person narrative is often implicit in procedural language, such as in headings ("[You] Archive client data") and imperative steps ("[You] Press the Start button").

16.4 [Theory] Perfecting Prose

Example: First Person	Example: Second Person	Example: Third Person
Pronouns: I, me, my, myself, we, our	Pronouns: You, your, yours, yourself, yourselves	Pronouns: He, him, his, himself, she, her, hers, herself, it, its, itself, they, them, their, theirs, themselves
"I pressed the Start button to start the engine."	"You must press the Start button to start the engine."	"He pressed the Start button to start the engine."

When to Use:

- When providing step-by-step instructions to guide the reader
- To create a sense of involvement and engagement in the content
- When a conversational tone is desired to simplify complex information

Chapter 17 Write Draft

Lead Writers: Steve Moss, Kieran Morgan

This chapter guides technical writers through the drafting stage. It introduces the iterative nature of writing, emphasizing the interplay between the writer, subject matter experts, and the product, process, or software system being documented. The chapter covers theoretical aspects by outlining key drafting stages, the write-review-update cycle, and the nuts and bolts of establishing version control. Practically, it instructs on preparing and writing the first draft, and engaging with subject matter experts through interviews and workshops.

 Who Should Read This

- Aspiring Technical Writers
- Beginner Technical Writers
- Cross-Domain Professionals

CONTENTS

17.1 Introduction 289
17.2 [Theory] The Stages of Drafting: First, Interim, and Final Drafts 290
17.3 [Theory] The Write–Review–Update Cycle 291
17.4 [Theory] Version Control 293
17.5 [Practice] Write First Draft 296

Chapter 17 Write Draft

PROCESS

Inputs	Write > Write Draft		Outputs
• Collected Information • Documentation Plan • Audience Profile or Persona	**Theory**	• The Stages of Drafting • The Write-Review-Update Cycle • Version Control	• First Draft Ready for Review • Updated Document Structure (e.g., Table of Contents)
	Practice	• Choose a Topic to Write • Prepare • Engage with Subject Matter Experts • Start Writing • Update Structure • Edit Draft	
	Samples	• NA	
	Templates	• Subject Matter Expert Interview Template • Subject Matter Expert Workshop Template	

17.1 Introduction

"Writing is easy. You just open a vein and bleed."

—*Unknown Author*[39]

This dramatic statement, often attributed to various authors, captures a certain truth about the emotional labor involved in crafting a piece of writing. However, when it comes to technical documentation, the process is less about emotional outpour and more about methodical precision.

We've spent ample time discussing preparation; now, let's dive into the act of writing itself. Writing is an iterative cycle of drafting, reviewing, editing, and polishing—and then doing it all over again, until your document is just right. This stage, writing the draft, is where you transform your detailed planning, collected information, and document structure (also known as the table of contents) into a series of drafts.

In this chapter, we focus on several related concepts: the stages of drafting, the write-review-update cycle, and establishing version control. We then conclude with practical steps for starting your first draft and templates to assist you in conducting subject matter interviews and workshops—essential steps in your preparation for writing.

If your documentation needs to include images, such as screenshots, photographs, or diagrams, and you're unsure about best practices, check out Chapter 18: Include Images.

17.2 [Theory] The Stages of Drafting: First, Interim, and Final Drafts

It's rare for a perfect first draft to emerge. In practice, most documents are written, reviewed, tested, and revised multiple times. This may seem like a waste of time for those new to writing, but it's an essential part of the process. The repeated cycles of review and revision improve your document's technical accuracy, enhance the quality and clarity of the writing, address gaps and weak spots, and allow for enhancements and updates.

Documents usually go through three stages of drafts—first draft, interim (or working) drafts, and final draft, as shown in the diagram below. This chapter focuses on the first draft stage. The following chapters, Chapter 19: Edit Drafts and Chapter 20: Review Draft, discuss the steps required to progress from a first draft to an interim and then a final draft, ready for approval and publication.

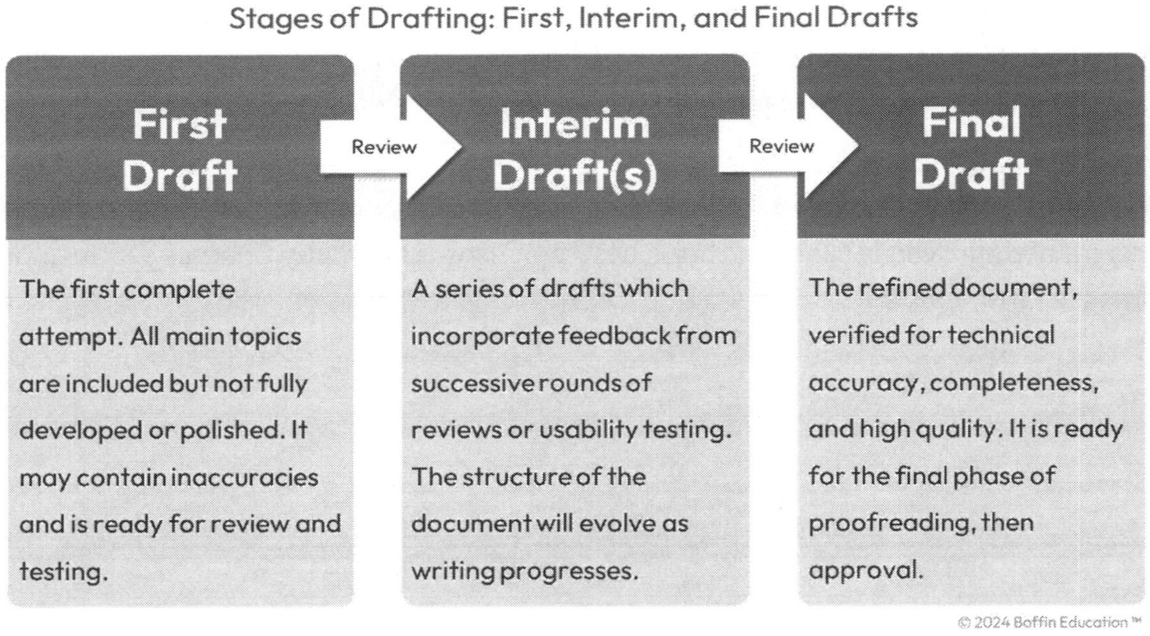

Stages of Drafting: First, Interim, and Final Drafts

First Draft	→ Review →	Interim Draft(s)	→ Review →	Final Draft
The first complete attempt. All main topics are included but not fully developed or polished. It may contain inaccuracies and is ready for review and testing.		A series of drafts which incorporate feedback from successive rounds of reviews or usability testing. The structure of the document will evolve as writing progresses.		The refined document, verified for technical accuracy, completeness, and high quality. It is ready for the final phase of proofreading, then approval.

© 2024 Boffin Education ™

Figure 41: Stages of Drafting: First, Interim, and Final Drafts

17.3 [Theory] The Write-Review-Update Cycle

Writing and reviewing is a highly iterative process. Drafts may go through any number of cycles of reviewing and updating until the final draft is ready for approval. The number of cycles, and their timing and duration, will be influenced by a smorgasbord of factors—the complexity of the subject matter, your experience and skill as a technical writer, the availability of reviewers, and more. Writing isn't an exact science!

The accompanying diagram depicts the cyclical nature of the write-review-update cycle. As you can see, it's the Interim Draft stage that introduces this uncertainty. This highly iterative phase, involving multiple cycles of review and update, is an unavoidable part of the process. No matter how good your first draft, it will always benefit from refinement by review. We discuss review in detail in Chapter 20: Review Draft.

Editing is your constant companion at every stage in drafting. Every time you write or update a draft, no matter what stage, it should always be edited—although the level of rigor in your editing will differ as you progress from the first through to the final draft, as we explain in Chapter 19: Edit Drafts. The cycle ends when the final draft is edited, proofed, and ready for approval.

Chapter 17 Write Draft

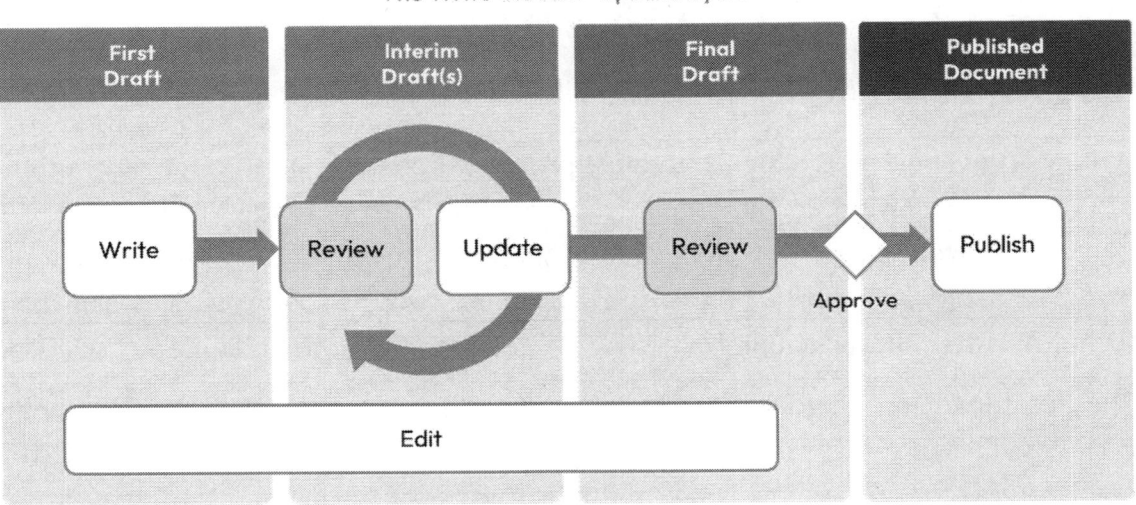

Figure 42: The Write-Review-Update Cycle

 Insight

Refining Words: Honing, Polishing, and… Sanding?!

Refining words is a craft akin to carpentry. Just as a carpenter sands down rough edges to reveal the smooth surface beneath, a writer continually hones and polishes their prose. This isn't the same as editing; it's a more subtle, continuous process of enhancement. With practice, this skill becomes second nature, integral to the act of writing itself.

17.4 [Theory] Version Control

Version control is the process of tracking and managing documentation as it's updated. It puts you in the conductor's seat, enabling you to orchestrate the write-review-update cycle and keeping what can be a complex process manageable. It's all part of the good housekeeping that any good technical writer should practice—and may be required under your organization's policies or standards such as ISO 9001.

Version control is often done automatically by content management systems. If not, there's a simple, tried-and-trusted system used by technical writers and knowledge workers alike: the version (or revision) number system.[40]

This method works by appending a numerical version number at the end of the document file name after every update, as shown in the diagram below. It's particularly helpful when the write-review-update process occurs via email, necessitating strict version control by the writer.

Chapter 17 Write Draft

Figure 43: Document Version Control

17.4 [Theory] Version Control

 What Does That Mean?

ISO 9001

ISO 9001 is the global standard for quality management systems. It's published by the International Organization for Standardization, a nongovernmental organization that develops international standards across a wide range of industries and sectors. This standard focuses on ensuring companies meet customer needs for their products and services. It also sets out requirements for document control, which we discuss in depth in Chapter 24: Publish.

Version

- **Meaning 1**: The specific identifier assigned to a document each time it is updated. This versioning allows for tracking the evolution of a document, as different versions (or "revisions") represent the document at various stages of its editing history.
- **Meaning 2**: A document's adaptation for different user groups while maintaining the same core content. For example, the same procedure document could exist in two versions: one in English and another in Spanish.

Version Control

A systematic process for tracking and managing different iterations of a document as it is revised and updated over time. This process ensures that all changes made to a document are recorded, enabling an audit trail of its evolution.

Tip

Implementing Version Control for Uncontrolled Content

Effective version control can be challenging to implement when developing topics in systems not designed for it. Topics may be reviewed in an online platform or a source code repository, which may not support the nuanced review cycle described here. In such cases, it might be practical to export online topics into Google Docs or Microsoft Word documents. This allows them to be tracked during the write-review-update process. However, this approach can lead to some messy housekeeping when it's time to update the content online at the conclusion of the review cycle.

17.5 [Practice] Write First Draft

You're now standing at the threshold of writing the first draft. This stage is where your planning and information gathering pay off. This section guides you through transforming your plans and research into a coherent, structured first draft that lays the foundation for what will eventually be a polished and professional document.

The main inputs for this stage are the information you collected in Part 4: Plan and the draft document structure (also known as the table of contents) you developed in Part 5: Design. As you write, you'll be guided by the knowledge of the purpose and audience of your document, which you defined when you wrote your documentation plan (see Chapter 10: Make a Plan). With every sentence you write, you'll apply the writing principles you learned in Chapter 16: Writing Principles.

This section guides you through the process of taking those inputs and developing that critical first draft.

17.5.1 Step 1: Choose a Topic to Write

Select a topic from your document structure that you mapped out in Chapter 13: Design Structure. It's best to choose a topic that you have plenty of information on already. This will allow you to quickly make good progress. The topic may be a high-priority task or one that you are familiar with from experimenting with the product.

17.5.2 Step 2: Prepare

Before writing the first draft, preparation is key. It's about gathering the information you need to construct a comprehensive and accurate first draft for the topic you're going to focus on initially. You should already have the background information you gathered in Chapter 9: Collect Information. This section is about bringing together the specific information you need to write your chosen topic.

Here are some common techniques technical writers use to prepare:

- **Engage with Subject Matter Experts**: Interviewing subject matter experts is an essential part of the preparation process. It involves detailed discussions, notetaking, and capturing screenshots where necessary. Early in the project, you might set up a series of meetings to progressively develop the document's content. When dealing with multiple subject matter experts on the same topic, workshops can be an efficient way to collect all their input simultaneously. We'll cover more on this in the following section.

- **Apply Your Knowledge**: If you're writing about a product you're familiar with—or are becoming familiar with through experimentation—you can use your knowledge to draft the document. Get hands-on with the product in a testing environment, such as a sandbox for software, to understand it better. This hands-on testing has the benefit of enabling you to personally verify the procedures you're writing about. This is critical for developing accurate procedures.

- **Use Legacy Documentation**: Start with what's already there. Existing documents can serve as a starting point for your new content. Depending on their quality, you might only need to update or edit parts of them. If they're outdated or inaccurate, you could be looking at a full rewrite. Be cautious with old material; sometimes it can lead to more issues than it solves. However, don't ignore potentially useful bits of knowledge tucked away in previous documents.

- **Delegate Writing Tasks to Others**: Getting others to write content for you can save time on internal documentation projects, especially when the writers are few—or just you!—and the scope is large. In some projects, writers act as both project manager and editor to a group of subject matter experts, assigning topics to the experts. If that's you, provide them with clear templates and well-written examples to guide their writing. Keep track of progress with a Status Tracker Template or Sample Documentation Project Schedule to ensure timely delivery, and be ready to send out reminders to keep everyone on track.

17.5.3 Step 3: Engage with Subject Matter Experts

In most documentation projects, your primary port of call for collecting information is subject matter experts. This is where you gain firsthand knowledge from those who know it best. You'll interview them, take notes, or perhaps even document as you go, capturing screenshots and noting down the sequence of steps on the fly. This information is a direct input, if not the source material, for your first draft. In this section, we'll discuss the two main approaches for engaging with subject matter experts: interviews and workshops.

17.5.3.1 Interviews

Interviews—either face-to-face or online—are a staple in the technical writer's toolkit. They're an excellent way of collecting information firsthand, whether it's about a product, service, or process. Good interviews don't just help you create accurate, high-quality documentation—they also help build rapport with colleagues.

Our Subject Matter Expert (SME) Interview Template explains how to prepare for the interview, including timing and duration, questions to ask, and a suggested sequence.

When to Use:

- **Exploring Complex Topics**: Interviews are ideal for unpacking complex or technical subjects, such as explaining a complex feature, validating the exact sequence of steps in a procedure, or capturing screenshots as a sequence of steps is demonstrated in a software application.

Seven Tips for Conducting Good Subject Matter Expert (SME) Interviews

1 Pre-Interview Preparation

Ensure a productive SME interview by understanding the topic, setting clear objectives for the interview, and sharing a set of questions beforehand.

2 Build Rapport

Start the interview with a comfortable tone, explain the purpose of the interview, and engage in active listening to establish a positive rapport with the SME.

3 Question Effectively

Start with broad questions, then narrow down to specific ones. Encourage detailed responses from the SME, and ask clarifying questions if anything isn't clear.

4 Manage the Interview

Keep the interview focused on the main topic, yet be flexible to open up new directions. Diligently take notes or recordings to capture any spontaneous insights.

5 Capture Technical Details

Request detailed step-by-step explanations and practical examples, using screen sharing to capture complex processes, screenshots, or functionalities.

6 Post-Interview

Summarize and validate key points with the SME post-interview, and don't hesitate to follow up with them to ensure accuracy in your draft.

7 Build Relationships

Foster a lasting collaborative relationship with the SME by expressing gratitude, keeping them informed on progress, and sharing the final document.

© 2024 Baffin Education ™

Figure 44: Seven Tips for Conducting Good Subject Matter Expert (SME) Interviews

> **Insight**
>
> **Use AI to Save Time in Notetaking and Transcribing**
>
> To save time, consider using AI tools such as ChatGPT to convert your rough notes into properly formatted minutes and actions. For more information on how to do so, see Chapter 8: Artificial Intelligence (AI) for Technical Writers.
>
> If a transcription of a recording is necessary and this functionality isn't built into your collaboration software, consider sending a recording to a transcription service such as Rev AI Speech to Text Transcription Service. Before you record, make sure to ask your subject matter expert for permission to do so.

17.5.3.2 Workshops

Workshops—either face-to-face or online—are an excellent way of gathering information from multiple subject matter experts simultaneously in a group setting. In a workshop, you'll play the role of facilitator, ensuring that each subject matter expert gets an opportunity to contribute and that discussions stay on point.

Our Subject Matter Expert (SME) Workshop Template explains how to prepare for a productive workshop, including timing and duration, questions to ask, and a suggested sequence.

When to Use:

- **Structuring and Planning**: When reviewing draft structures (tables of contents, lists of topics, and so on) for documentation or brainstorming which deliverables are required for a new project
- **Getting Consensus on Processes or Procedures**: When reviewing processes or procedures, typically where there's some confusion about the correct sequence or a process hasn't been fully mapped out yet

Never try to finesse the wording of documents in a workshop. People will spend hours agonizing over a single sentence given the chance!

> **Tip**
>
> **Building Bridges: The Art of Collaborating**
>
> Building strong relationships with subject matter experts is important for both your work and your career. A good rapport leads to better access, quality feedback, and improved documentation. For subject matter experts, it ensures their work is understood and supported throughout the development process.
>
> To foster these relationships, writers should engage as full participants in the process by attending and contributing to meetings, respect subject matter experts' time by being accommodating, and offer help and insights to avoid a one-sided dynamic where the writer only takes information without contributing anything of value in return.
>
> Remember, effective collaboration is about creating a win-win outcome. When done right, it will expand your professional network with an ever-growing group of professionals who hold you in high esteem—and may well open up future job opportunities as time goes by.

17.5.4 Step 4: Start Writing

Now it's time to write. As technical documentation is usually task focused, a good place to start is by writing the procedural steps. That will lead you to other information that you'll need to complete the topic. This may include concepts necessary to understand the task, specific tools required to execute it, or other topics that are downstream or upstream of it. You may also find that there are what-ifs—conditional steps that need to be included, such as a failure mode that needs documenting or cross-references to troubleshooting topics.

As you work through your notes and incorporate them into your draft, you'll come across chunks of information that you'll want to include—enough for a sentence, table, diagram, paragraph, or perhaps more. Add these chunks into your draft, integrating them into the structure so that they don't interrupt the flow.

In your notes, keep track of information that you have already incorporated by marking it, for example, with strikethrough or a different color. This "mosaic" approach will let you integrate information from your notes into your draft structure without causing duplication—which you want to avoid.

 Tip

Write Introductions and Overviews Last

Although they always appear first in the table of contents, don't write introductions or overviews first. You won't have the necessary knowledge of the product, feature, or process at this stage. Only as you close in on the final draft stage will you have built your knowledge to the point where you can finally write the overview. To save time, you can use AI to help you do so—see Chapter 8: Artificial Intelligence (AI) for Technical Writers.

 Insight

For a Fresh Perspective, Sleep on It

After completing your first draft, set it aside for a day or two before revisiting it. This pause allows you to gain a fresh perspective on what you've written. When you return to your draft, you may discover that details previously overlooked now leap out at you, enabling you to easily identify areas that need refinement or clarification.

17.5.5 Step 5: Update Structure

As you write, you'll discover that your document's structure will evolve. This is completely normal—it's the result of building your knowledge about whatever you're documenting, knowledge you didn't have at the start of your project. Even a technical writer with deep knowledge of the subject will find that things evolve in ways they hadn't anticipated as they progress. No one has perfect knowledge!

Don't be surprised if you find that:

- Topics blow out into chapters—and chapters condense into topics.
- New topics are discovered that need to be added to chapters.
- New chapters are required to encompass topics you hadn't considered originally.
- The ordering of information in the document structure has changed.
- Appendices are now required for reference material.

17.5.6 Step 6: Edit Draft

Finally, you'll need to edit your first draft so that it's polished enough for a subject matter expert review. Although your first draft doesn't need to be perfect—and may contain things such as placeholders for information you haven't yet gathered—you don't want to send out a draft full of obvious mistakes in spelling, grammar, and punctuation. While it may save time, it could lead others to question your writing skills and your attention to detail.

In the next chapter, Chapter 19: Edit Drafts, we focus on the editing process and explain how to apply the appropriate level of editing to refine your document at the different stages of the write-review-update cycle.

Chapter 18 Include Images

Lead Writers: Swapnil Ogale, Kieran Morgan | **Peer Reviewer/s**: Castella Arthur | **Expert Reviewer/s**: John Sweller, Deirdre Wilson

This chapter outlines the importance of using visuals like screenshots and diagrams in technical documentation. It introduces two key theories—multimedia learning and cognitive load theory—to explain how the human brain processes visual and textual data. It provides practical tips on selecting and preparing visuals and includes a checklist for effective implementation. Overall, this chapter aims to equip technical writers with both the theoretical understanding and practical tools needed to enhance documentation through imagery.

 Who Should Read This

- Aspiring Technical Writers
- Beginner Technical Writers
- Cross-Domain Professionals

CONTENTS

18.1 Introduction 307
18.2 [Theory] Multimedia Learning: Your Brain Has a Dual-Core Processor 307
18.3 [Theory] Cognitive Load Theory 310
18.4 [Theory] Bringing It All Together: Cognitive Load and Technical Documentation 314
18.5 [Practice] Crop, Capture, and Caption Images 315
18.6 [Practice] Working with Graphic Designers 322
18.7 [Practice] Translation Considerations 323

Chapter 18 Include Images

PROCESS

Inputs	Design > Include Images		Outputs
• System Access • Access to Subject Matter Experts • Style Guide • Screen Capture and Image Editing Tools	**Theory**	• Multimedia Learning • Cognitive Load Theory	• Technical Documentation with Embedded Images
	Practice	• Crop, Capture, and Caption Images • Five Rules for Using Images in Technical Documentation • Working with Graphic Designers • Translation Considerations	
	Examples	• Example of a Captioned Image	
	Templates	• Checklist for Using Images in Technical Documentation	

18.1 Introduction

This chapter explains the theory and practice of using visuals in technical documentation. Most tech writers are familiar with using screenshots and visuals, but there are techniques to follow to get it right. Beneath the apparent simplicity of a screenshot or diagram lie some powerful concepts in human psychology and how your brain processes information. You've no doubt seen lists of dos and don'ts in information design, but why are these considered best practices? This chapter will explain the theories behind the practice and give you strategies to apply in your documentation projects.

The first theory is multimedia learning, which is akin to saying your brain has two processors, similar to a computer. These work in parallel to process images and text more efficiently. The next theory is cognitive load theory, which explains how your brain's processors convert sensory input into knowledge—and how that process can get bogged down by unnecessary, "extraneous" information. Finally, we talk about some extremely practical strategies you can use to put these theories into action.

18.2 [Theory] Multimedia Learning: Your Brain Has a Dual-Core Processor

Have you ever heard of a dual-core processor? Most of us have. If you've bought a computer anytime since the turn of the century, you've likely had dual-core advertising claims hurled at you like confetti. Thanks to the marketing folks, we now have a convenient metaphor to explain how multimedia learning works in your brain.

Chapter 18 Include Images

Let's imagine that your brain is a computer you're considering purchasing. The advertisement highlights its "dual-core processor." What does this mean? One processor is dedicated to images, and the other to language (including verbal, text, and sign language). By employing both "processors" concurrently—in parallel—your brain-computer can process information almost twice as efficiently. This foundational concept in instructional design is known as "multimedia learning theory."[41] According to this theory, using words and images in tandem makes learning, on average, 89 percent more effective than using words alone.[42]

18.2 [Theory] Multimedia Learning: Your Brain Has a Dual-Core Processor

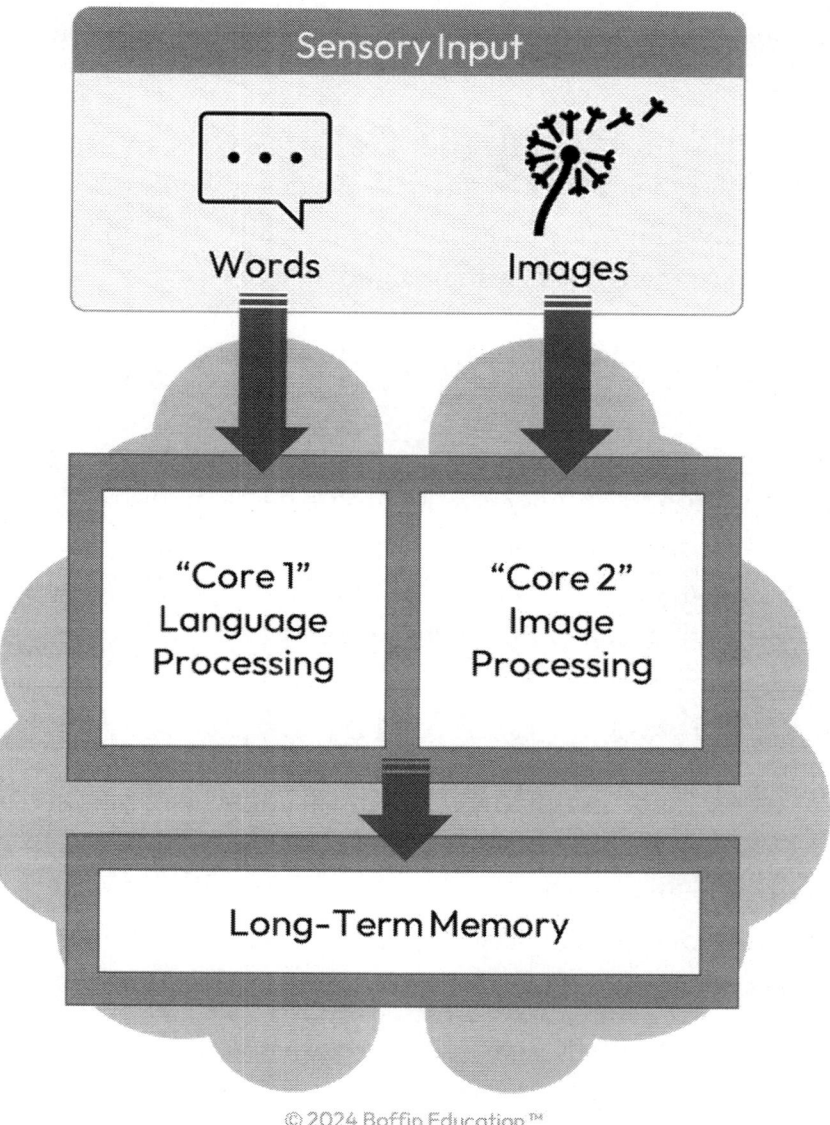

Figure 45: Your Brain's Dual-Core Processor

Of course, you have many more inputs than that. Your senses also include taste, smell, movement, and so on. However, for the purposes of technical writing—even if you're the Picasso of your craft—your technical documentation palette is usually limited to language and images.

For technical writers, the implications of this theory are clear and something that every experienced technical writer already intuitively knows. Using appropriately selected and well-captioned images alongside text will make your technical documentation much more effective for your audience.

18.3 [Theory] Cognitive Load Theory

Cognitive load theory explains how our brains process and manage new information. We're going to dive into this theory in a little more detail in this section. If you want to go straight to the practical applications, feel free to skip this part. But if you want to learn how people process information and how this powerful concept is relevant to the visuals in technical documentation, then read on.

18.3.1 Limited Processing Capacity

We all know the feeling of information overload—that ache behind the eyeballs when we know we've hit our limit. It's often accompanied by the fear that if someone asks us what we've just learned, we'll draw a complete blank. This isn't just the result of a lack of caffeine or too many sleepless nights spent cramming for exams; it represents a fundamental limit on your brain's ability to absorb information.

Why is this? It has to do with how your brain processes information. When you receive input from the outside world, it undergoes processing in your brain and is ultimately stored in your long-term memory, as shown in the diagram below.

18.3 [Theory] Cognitive Load Theory

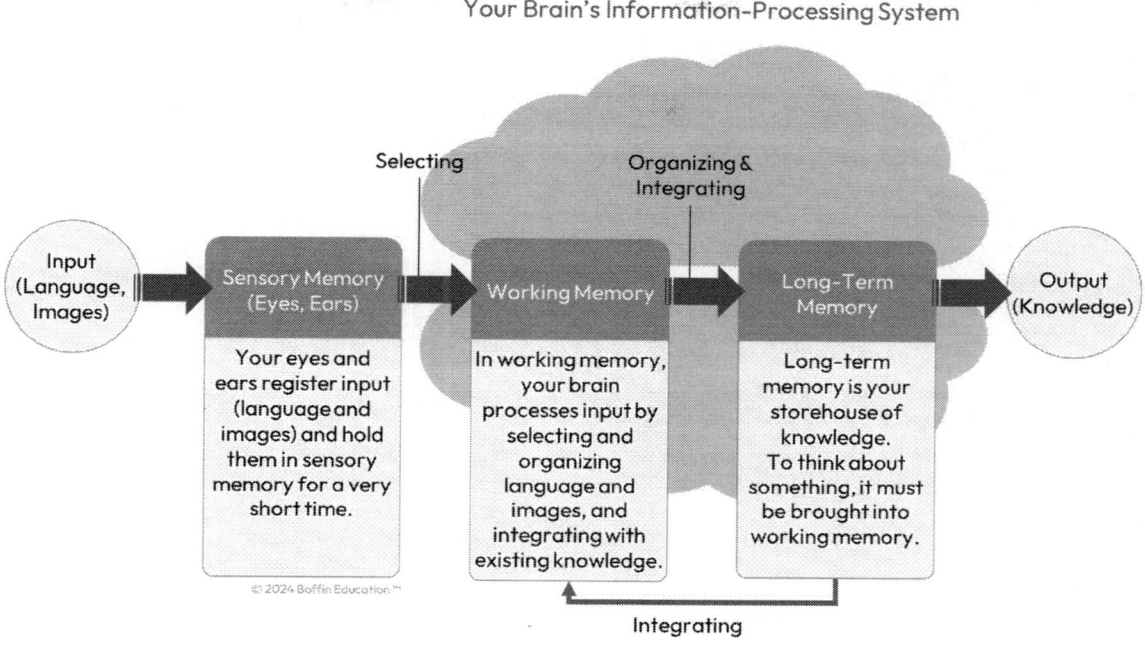

Figure 46: Your Brain's Information-Processing System

So, what's the problem? The issue lies in the working memory, which has a very limited capacity, compared with the world around you, which can offer an unlimited number of inputs. In fact, most people's brains can only hold up to four chunks of information at once,[43] and possibly even as few as two or three.[44]

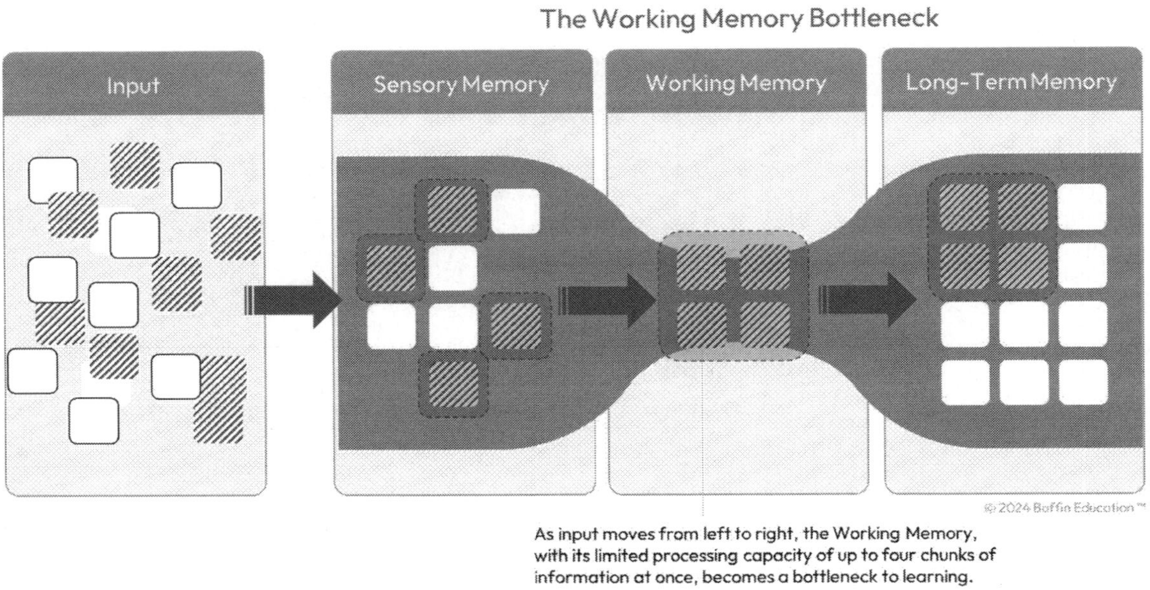

Chapter 18 Include Images

The Working Memory Bottleneck

As input moves from left to right, the Working Memory, with its limited processing capacity of up to four chunks of information at once, becomes a bottleneck to learning.

Figure 47: The Working Memory Bottleneck

18.3.2 Good Processing vs. Bad Processing

When was the last time you searched for information on how to carry out a task only to find yourself filtering out irrelevant information or trying to decipher confusingly worded or ambiguous instructions?

It's a headache, isn't it? That's because filtering and parsing different sources of information put a lot of strain on our limited working memory. This is the unnecessary extrinsic processing our brains have to do before we can focus on the vital intrinsic information.

If only we had perfectly tailored information. Imagine how much easier our lives would be! Your job as a technical writer is to help eliminate extrinsic information for your audience so they can focus on the important, intrinsic information.

The types of processing load you need to be aware of as a technical writer are:

1. **Intrinsic Load**: The effort involved in learning relevant, or intrinsic, information. This is "good" processing. For example, learning a concise sequence of well-written instructions on how to crop an image using an image editing tool should involve a fairly low intrinsic load as it's a straightforward procedure.

2. **Extrinsic Load**: The effort involved in filtering out irrelevant, or extrinsic, information. This is "bad" processing. For example, searching and filtering multiple hits in a knowledge base for the best information on how to crop an image, or trying to decipher poorly written, text-heavy instructions that go on and on and don't include images of key functions.

 What Does That Mean?

Intrinsic Load

The effort involved in learning relevant, or intrinsic, information, as derived from the field of cognitive psychology.

Extrinsic Load

The effort involved in filtering out irrelevant, or extrinsic, information, as per the field of cognitive psychology.

18.4 [Theory] Bringing It All Together: Cognitive Load and Technical Documentation

Now you understand that your brain operates much like a dual-core processor—a capability you can leverage as a technical writer. You're also aware that your audience's working memory can easily become overloaded with information, hindering their ability to process essential intrinsic details. This problem is further exacerbated when your instructions are cluttered with unnecessary extraneous load. As a technical writer, it's your responsibility to use this knowledge to make your instructions more effective.

The most common visuals you'll find in technical documentation include:

- **Screenshots** of a software system's user interface, such as for a user guide. Also known as screen captures.
- **Diagrams** of the components of a physical device, as in a hardware user manual.
- **Images**, such as diagrams, charts, and infographics, for documents like reports.
- **Photographs** of equipment in the field, such as for a procedure.

In this chapter, we'll discuss general principles for incorporating images into technical documentation, as well as specific strategies for dealing with these different image types.

18.5 [Practice] Crop, Capture, and Caption Images

In technical documentation, images are essential tools for conveying information effectively. The right visual can demystify complex ideas, guide users through detailed procedures, and boost overall comprehension. This section draws upon the theories outlined previously to offer practical strategies that enhance understanding of technical content, aiding users in efficiently completing tasks.

 Tip

Check Your Style Guide Before Including Visuals

Style guides often contain specific guidance on the use of visuals in documentation, ensuring alignment with your organization's branding. Uniformity in screenshots and other visuals, regardless of who creates them, is crucial for consistency. Before integrating visuals into your technical documents, review your style guide. It may have stringent rules about their use.

If your style guide lacks specific guidelines on screenshots, consider tailoring the items from our Checklist for Using Images in Technical Documentation as a foundation for your own style guide.

Ensure your style guide details stylistic elements such as preferred line colors, thicknesses, and callout fonts for screenshots. Establishing these elements clearly in writing promotes consistency and assists others in adhering to standard guidelines.

18.5.1 Five Rules for Using Images in Technical Documentation

The power of visuals lies not just in their use, but in their thoughtful implementation. We've distilled a set of best practices from industry that you can use to maximize the usefulness of visuals in your documents, regardless of their format.

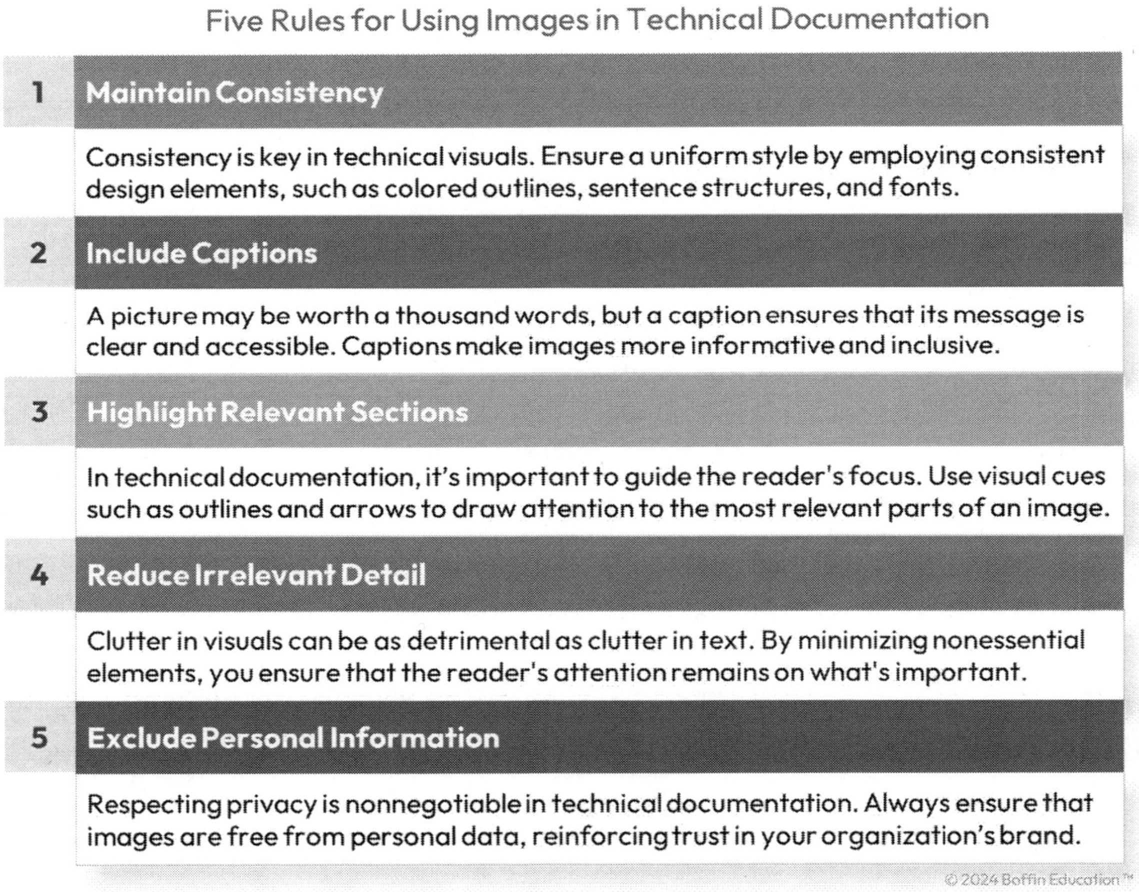

Five Rules for Using Images in Technical Documentation

#	Rule	Description
1	**Maintain Consistency**	Consistency is key in technical visuals. Ensure a uniform style by employing consistent design elements, such as colored outlines, sentence structures, and fonts.
2	**Include Captions**	A picture may be worth a thousand words, but a caption ensures that its message is clear and accessible. Captions make images more informative and inclusive.
3	**Highlight Relevant Sections**	In technical documentation, it's important to guide the reader's focus. Use visual cues such as outlines and arrows to draw attention to the most relevant parts of an image.
4	**Reduce Irrelevant Detail**	Clutter in visuals can be as detrimental as clutter in text. By minimizing nonessential elements, you ensure that the reader's attention remains on what's important.
5	**Exclude Personal Information**	Respecting privacy is nonnegotiable in technical documentation. Always ensure that images are free from personal data, reinforcing trust in your organization's brand.

© 2024 Boffin Education™

Figure 48: Five Rules for Using Images in Technical Documentation

Example

Here's a simple example of a captioned image, using typical best-practice elements such as outlined boxes and contrasting colors and font sizes. This is from Chapter 22: Translation Theory.

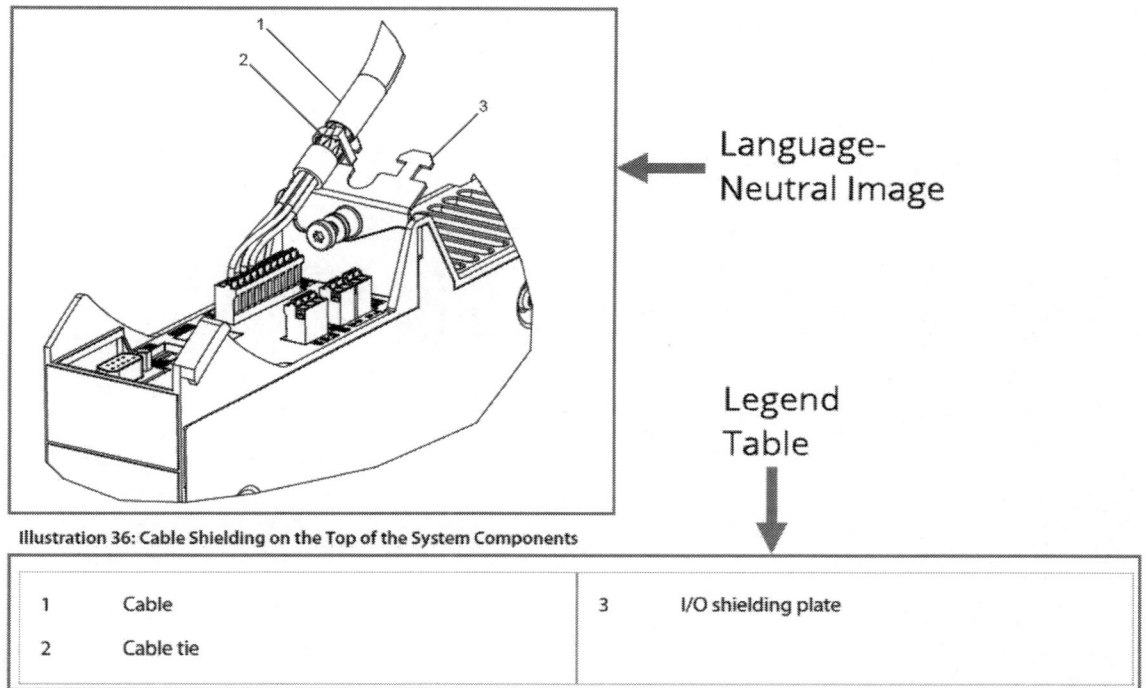

Figure 49: Example of a Captioned Image

 **Note**

Ensure Copyright Compliance

Always be vigilant about copyright when sourcing visuals. Remember, citing a source does not necessarily equate to having permission for use.

18.5.2 Screenshots

If you're a technical writer working on software documentation, you will be involved in taking screenshots and adding them to your documentation. If you're thinking to yourself, "How many screenshots do I include?", the answer may range from "None" to "As many as possible." The quantity of screenshots you use is determined by many factors. While screenshots can improve the quality of your technical writing and provide a visual medium for users to complete their tasks, they may be a hindrance if not used properly.

When to Use Screenshots in Technical Documentation

Use More...	Limit...	Don't Use...
✓ If screenshots help explain a complicated procedure, function, or user interface better than words alone. ✓ If screenshots demonstrate a new feature or change in functionality to the visual user interface. ✓ If screenshots help users validate that they've correctly performed a procedure. ✓ If the use of screenshots aligns with your audience's or stakeholders' expectations.	✓ If the user interface changes frequently, as screenshots will become quickly outdated. ✓ If screenshots need to be translated into different languages or sourced in different languages. ✓ If customizable user interfaces make it impossible to capture all possible software configurations.	✓ For code samples. Consider more helpful options such as code blocks that allow users to copy and paste the code without having to manually type it out. ✓ For commonplace graphical elements such as progress bars, confirmation dialog boxes, welcome screens, or standard buttons (OK, Cancel, Yes, No). ✓ If it takes more time for your audience to interpret the visual than a simple written instruction.

© 2024 Boffin Education™

Figure 50: When to Use Screenshots in Technical Documentation

To ensure high-quality screenshots for technical documentation, it's crucial to have a well-organized process in place. The sequence below outlines how you can systematically source and capture screenshots.

18.5 [Practice] Crop, Capture, and Caption Images

1. **Prepare**: Verify that you have access to the relevant software or user interface you wish to document. Determine if you'll need administrator-level access to explore all the software's functionalities. For more information, see Chapter 9: Collect Information (Secure Access to Hardware and Systems).

2. **Consult the Software Team**: The software team is often busy with development tasks. Ensure that taking screenshots won't disrupt their workflow.

3. **Check the Style Guide**: Look for specific guidelines for taking screenshots. This helps in maintaining a consistent quality across all captures. If your style guide doesn't have any guidelines, use the Checklist for Using Images in Technical Documentation. You should also look for templates or best-practice examples that you can easily turn into guidelines.

4. **Select Tools**: Identify the screen capture tool you'll be using. Most operating systems, including Windows, Linux, and Mac, have built-in tools for basic screen capturing. For more efficient image editing, capturing, and captioning, many technical writers use software like SnagIt.

5. **Capture**: Start capturing the screenshots, adhering to the guidelines you've set. Make sure the images are clear, focused, and relevant to the topic by following our Checklist for Using Images in Technical Documentation.

6. **Enhance Images**: After capturing, use image editing tools to crop, enhance, and manipulate the screenshots as needed.

7. **Store and Organize**: Store the screenshots in an organized manner, making it easier for future updates or revisions.

By following this sequence, you not only facilitate the process of capturing screenshots but also improve the overall quality of your technical documentation.

> **Tip**
>
> **Tailoring Your Captures to the User's Device**
>
> Keep in mind how users will interact with the software. Will they be using a desktop computer or a mobile phone? This distinction is crucial, as screenshots will appear differently in these varied environments. Tailoring your captures to the user's device improves the clarity of your documentation.

18.5.3 Diagrams and Photographs

If you're a technical writer working on hardware or procedural documentation, you'll need to understand how to source, capture, and caption diagrams (such as flowcharts, hardware component diagrams, and infographics) and photographs (for example, of equipment in the field). Hardware and procedural documentation may even involve a combination of screenshots and visuals, for instance if a task involves an element of interaction between software and a physical device.

18.5 [Practice] Crop, Capture, and Caption Images

When to Use Diagrams and Photographs in Technical Documentation

Use More...	Limit...	Don't Use...
✓ When diagrams and flowcharts distill complicated concepts and processes into more digestible information than words alone. ✓ When photographs verify the physical setup or indicate how to conduct a successful equipment installation. ✓ Where diagrams clarify different hardware components or the sequence required to assemble or maintain them.	✓ If the subject matter changes frequently (e.g., 3D renders of a prototype product are continually updated), as diagrams can quickly become outdated. ✓ If diagrams need to be translated into different languages or sourced in different languages.	✓ When very simple instructions can be more efficiently conveyed through text alone.

© 2024 Boffin Education ™

Figure 51: When to Use Diagrams and Photographs in Technical Documentation

The steps below describe how to source and create high-quality diagrams and photographs for technical documentation.

1. **Prepare**: Define the purpose and context where the diagram or photograph will be used. For equipment photographs, make sure you have access to the hardware or field location.

2. **Consult the Team**: Discuss with relevant teams (hardware engineers, field operators, and others) to ensure that capturing visuals won't disrupt workflows.

3. **Check the Style Guide**: Consult your style guide for visual elements other than screenshots. If it's lacking, develop guidelines by referring to the Checklist for Using Images in Technical Documentation. You should also look for templates or best-practice examples that you can easily turn into guidelines.

4. **Select Tools**: Choose appropriate software for diagram creation, such as Visio, Lucidchart, or specialized engineering software. For photographs, ensure you have a high-quality camera.

5. **Capture or Create**: Either capture the photograph or create the diagram according to the established guidelines. Ensure clarity, focus, and relevance.

6. **Enhance Images**: Use image editing tools for cropping, color correction, or adding labels and callouts to photographs. For diagrams, ensure that lines are clear and text is legible.

7. **Store and Organize**: Store these visuals in an organized manner to facilitate future updates or revisions.

18.6 [Practice] Working with Graphic Designers

If you're collaborating with a graphic designer—perhaps on a project where your copy and visuals will be transformed into a designed or printed manual or report—there are key considerations to keep in mind for a smooth process. By adhering to these practices, you can enhance the visual consistency of your technical documentation when collaborating with graphic designers and reduce the likelihood of rework.

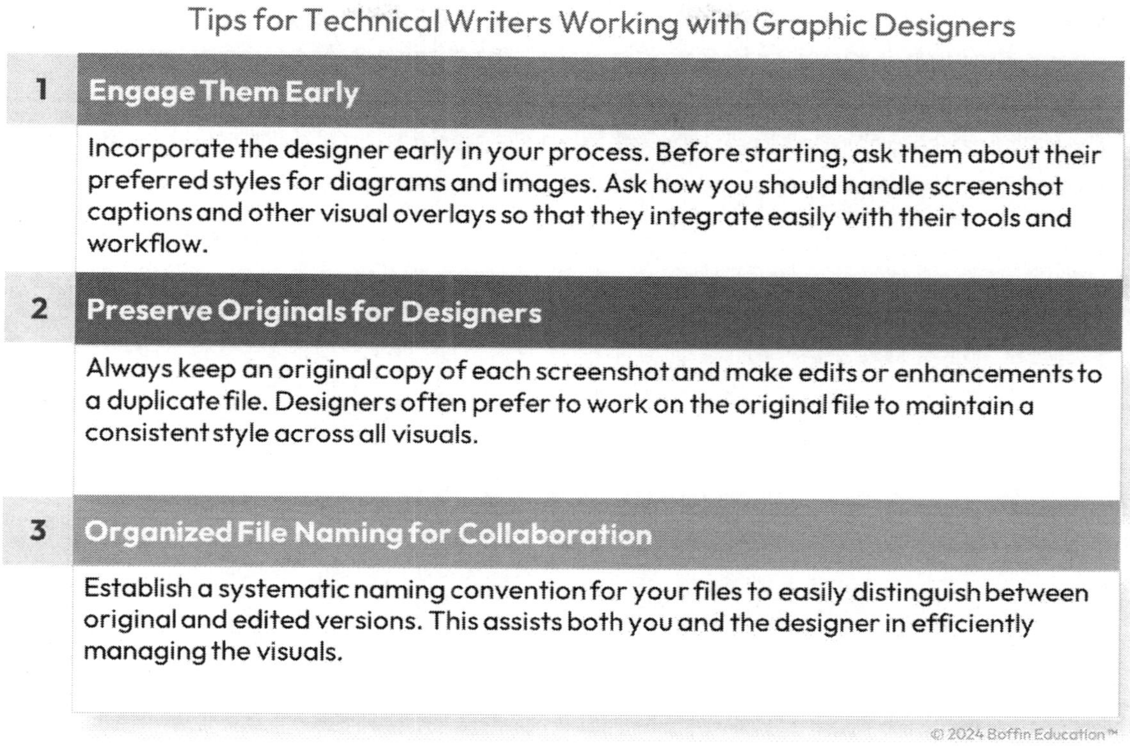

Tips for Technical Writers Working with Graphic Designers

1 Engage Them Early

Incorporate the designer early in your process. Before starting, ask them about their preferred styles for diagrams and images. Ask how you should handle screenshot captions and other visual overlays so that they integrate easily with their tools and workflow.

2 Preserve Originals for Designers

Always keep an original copy of each screenshot and make edits or enhancements to a duplicate file. Designers often prefer to work on the original file to maintain a consistent style across all visuals.

3 Organized File Naming for Collaboration

Establish a systematic naming convention for your files to easily distinguish between original and edited versions. This assists both you and the designer in efficiently managing the visuals.

© 2024 Boffin Education™

Figure 52: Tips for Technical Writers Working with Graphic Designers

18.7 [Practice] Translation Considerations

If your technical documentation will be translated, there are some considerations you should take into account when you're including visuals:

- **Sourcing Visuals in Different Languages**: If you're creating visuals that include user interface text (like software), you'll need to consider where you'll source visuals for foreign-language translations. You may not be able to take them yourself, as the use of another language will make it complicated for you to navigate the software. If this is the case, talk to the software development team early to make sure they can arrange for someone to provide the screenshots in the languages required when needed.

- **Use of Numbered Callouts**: If the software is in a single language—or you're documenting hardware or field equipment—then it is better to use numbered callouts with an explanatory table below the visual. This will reduce translation costs, because only the text in the table needs translating. The translation partner won't have to edit graphics, which will result in additional costs.

For more information about translating technical documentation, see Chapter 22: Translation Theory and Chapter 23: Translation Practice.

Part 7 Edit

Refining technical writing through precise editing for clearer communication.

Chapter 19 Edit Drafts

Lead Writer: Kieran Morgan | **Peer Reviewer/s**: Steve Moss, Felicity Brand

This chapter focuses on the editing stage and details the levels and tools of editing in technical writing. It outlines the traditional levels of editing while adding specific ones tailored for technical writing. The chapter emphasizes the iterative nature of editing and the importance of various inputs, such as style guides and dictionaries. It also discusses the use of artificial intelligence in editing and provides practical steps for the editing process to ensure that documents are clear, consistent, and accurate.

 Who Should Read This

- Aspiring Technical Writers
- Beginner Technical Writers
- Cross-Domain Professionals

CONTENTS

19.1 Introduction 329
19.2 [Theory] Levels of Editing 331
19.3 [Theory] Editing Tools and Inputs 338
19.4 [Practice] Edit Draft 342

Chapter 19 Edit Drafts

PROCESS

Inputs	Edit > Edit Drafts		Outputs
<ul><li>Style Guide</li><li>House Style Manual</li><li>Technical Writing Style Guide</li><li>Industry-Specific Style Guides</li><li>Brand Guidelines</li><li>Dictionary</li><li>Access to Internet</li></ul>	**Theory**	<ul><li>Levels of Editing</li><li>Editing Tools and Inputs</li></ul>	<ul><li>Edited Drafts</li></ul>
	Practice	<ul><li>Determine Level of Edit Needed</li><li>Apply Inputs</li><li>Utilize Tools</li><li>Review Draft</li></ul>	
	Examples	<ul><li>N/A</li></ul>	
	Templates	<ul><li>Editing Checklist Template</li><li>Editing Sheet Template</li></ul>	

19.1 Introduction

So, you think you've written the perfect first draft. You've diligently applied the principles outlined in Chapter 16: Writing Principles, and followed the guidance in Chapter 17: Write Draft. You're eager to send your first draft for review, allowing subject matter experts to validate its accuracy and provide feedback on its effectiveness.

But wait! Don't rush to send it off without a thorough edit, or you risk looking unprofessional. An unedited document riddled with errors can portray you as careless or, worse, uninformed. It could waste others' time picking up mistakes you should have already attended to. As technical writers, we're expected to exhibit expertise in both writing and editing, and abundant attention to detail. Demonstrate your skills by sending a well-edited document for review to showcase your commitment to your craft.

What exactly is editing? It's the time-honored process of refining a document to reach its final draft stage, when it's ready for publication. Editing typically involves multiple rounds of review at varying levels. The goal of editing is to ensure that your document is as clear, consistent, and accurate as possible. As discussed in the previous chapter, Chapter 17: Write Draft, editing is an integral part of the write-review-update cycle. Each update to a document requires editing to maintain the quality and structural integrity of the overall work.

Editing relies on two key tools: the Editing Checklist Template and the Editing Sheet Template, as discussed in Chapter 19: Edit Drafts: Editing Tools. It also depends on various inputs like style guides and dictionaries, which we explore in Chapter 19: Edit Drafts: Editing Inputs. The practice of editing itself demands a keen eye for detail and a nuanced understanding of meaning, as explained in Chapter 19: Edit Drafts: Edit Draft.

The stage that follows editing is the review process, which we'll explore in Chapter 20: Review Draft.

Tip

Use Artificial Intelligence (AI) to Save Time in Editing

To save time in editing these days, you can use AI tools, AI-assisted checkers like Grammarly, or "prose linters" if you're working in a docs-as-code environment. Even ChatGPT does a decent job of editing.

Why bother learning about editing? Isn't it somewhat passé now? Well, no. There are aspects of editing that AI won't help you with, such as validation, compliance checking, and layout and formatting. These stages are critical for technical writers.

It's also important that you understand the concepts and terminology of editing. Doing so will enable you to work more effectively with AI tools. You can incorporate the levels of editing into your prompts, and a tool like ChatGPT will understand the difference between a structural edit and proofreading, for example.

Finally, remember that these tools are far from perfect. You still need to carefully review their output. See Chapter 8: Artificial Intelligence (AI) for Technical Writers for more information.

 What Does That Mean?

Prose Linter

A tool or software application designed to analyze written text for style, grammar, syntax, and sometimes even content consistency. These tools go beyond spell checkers by offering a more in-depth analysis of the text. The concept is borrowed from the world of programming, where a "linter" is a tool that analyzes source code to flag programming errors.

19.2 [Theory] Levels of Editing

Editing is traditionally divided into three levels: structural editing, copy editing, and proofreading.[45] To these levels, we've added several others that we consider essential for technical writers: validation, compliance checking, and layout and format checking. The addition of these levels ensures that your document isn't just well-written; it meets all the requirements generally considered essential for publication as a technical document.

As you develop and mature your draft document, you'll find yourself moving from left to right through the levels of editing. You'll notice that there's some overlap between the levels; for example, both proofreading and copy editing involve spelling and grammar checks. This is intentional. Editing is an iterative process involving numerous back-and-forth passes, and there is no hard-and-fast rule on exactly what falls into each level of editing or even how many levels exist.

The levels of editing framework we've defined has been tailored specifically for technical writing while drawing on traditional, well-understood editing concepts and terminology. It utilizes two tools at various levels: the Editing Checklist Template and the Editing Sheet Template, which we'll discuss next.

Chapter 19 Edit Drafts

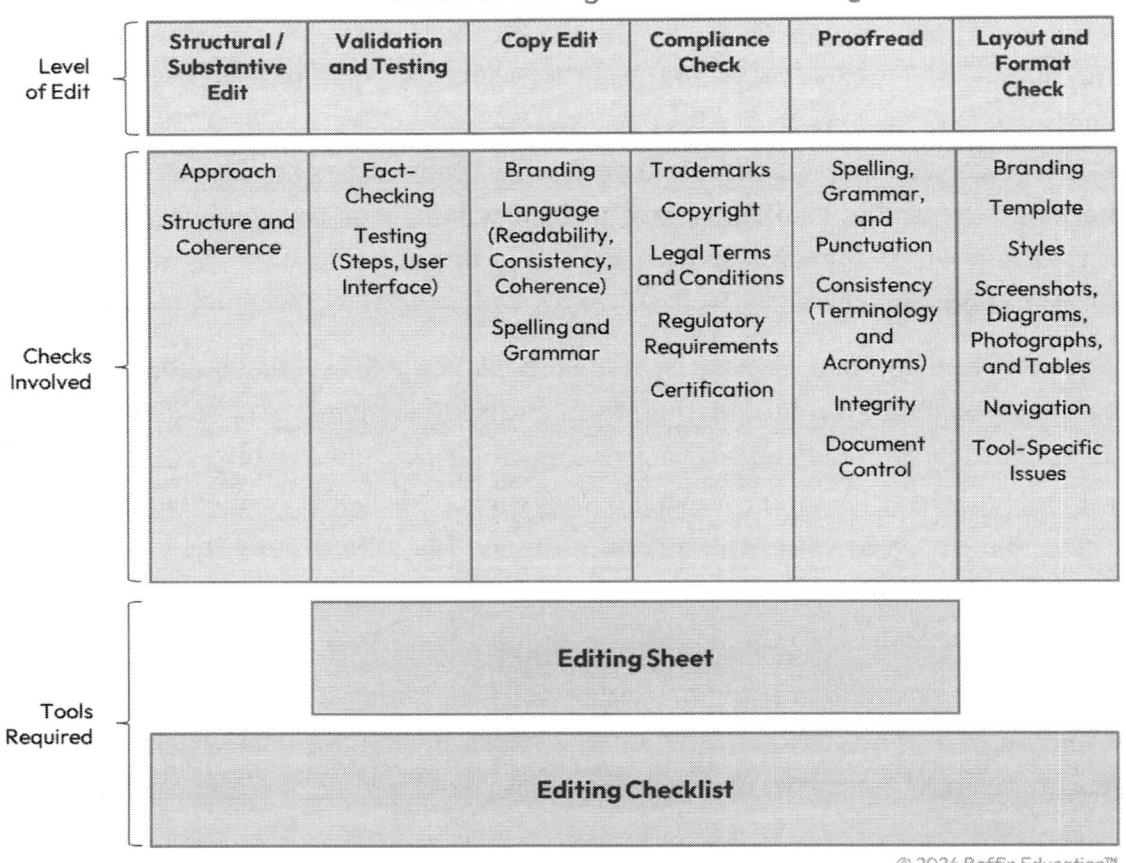

Figure 53: Levels of Editing in Technical Writing

19.2 [Theory] Levels of Editing

 Insight

Tech Writers Wear Many Hats

In most small- to medium-sized organizations, the technical writer often assumes multiple responsibilities or "hats," such as writer, editor, and designer. In contrast, very large organizations may support specialized or dedicated roles, such as technical editors and designers who handle the layout. You never know where you'll end up as a writer, so it's useful to develop a well-rounded skill set.

19.2.1 Structural or Substantive Editing

Structural or substantive editing is an in-depth analysis of a document's approach, structure, and coherence. It evaluates whether a document meets its goals and aligns the editing to these objectives. Structural editing may involve a substantial reorganization of the document or a change in approach.

It focuses on:

- **Approach**: The writer's general approach to solving the challenges presented in the documentation plan is reviewed. This involves placing the document in context and asking challenging questions such as: Is this document fit for purpose? Is it written appropriately for the defined audiences, and does it adequately address the issues they face? Is the use of language, including tone and narrative point of view, appropriate? Are there any weak points? Does the document succeed overall?

- **Structure and Coherence**: This part of the review ensures that the document progresses logically. It checks that context is provided in the appropriate places (such as overviews, introductions) and that each section and topic flows naturally to the next. Navigational elements and components, such as tables of contents, indexes, and appendices, should be considered, as well as structural elements within the content, like cross-references, hyperlinks, callouts, and the use of tables.

- **Completeness and Relevance**: It's important to also check for completeness (everything necessary has been covered) and relevance (all included information is required in the document).

When to Use: Early in the writing process to set a strong foundation for the document. Midstream in the writing process when integrating or reworking new content into an existing document.

 Insight

When Substantive Editing Becomes a Rewrite

Sometimes you'll find yourself going well beyond substantive editing. You may discover that the material you're working from is so unfit for its intended purpose that you're rewriting every sentence, resulting in an end product that bears almost no resemblance to the original. This scenario is common. It can occur when subject matter experts with varying writing skills hack together a document, leaving it to you to refine.

This presents a unique challenge for you as the writer to use your writing skills to create a coherent document while preserving the original's meaning. In such cases, you'll be drawing once again on the writing skills we discussed in Part 6: Write to collaborate closely with the subject matter experts to ensure your revisions don't lose any vital details.

19.2.2 Validation and Testing

Validation and testing is an essential step in ensuring the accuracy of technical documentation, as discussed in detail in Chapter 21: Validate and Test Information. Consider it non-negotiable!

To summarize, it involves two aspects:

- **Factchecking**: This process involves verifying technical facts against specifications data and reference documentation, or with key subject matter experts. For instance, you might confirm that a range of values described in a product's user guide matches those in the product specifications.

- **Testing**: During testing, the writer attempts to follow the steps in the document to verify that they produce the described result or works alongside a subject matter expert to do so. This also includes validating that any text or images from the product's user interface reproduced in the documentation match the product exactly.

When to Use: Shortly after a topic is written. This step ensures the technical accuracy and practical applicability of the content, which is essential before proceeding to more detailed editing stages and review.

19.2.3 Copy Editing

Copy editing is an in-depth review of a document's language, spelling, and grammar. The emphasis is on adding value by analyzing the use of language, its effectiveness for the target audience, and alignment with your organization's brand guidelines.

In copy editing, language is checked for readability, consistency, and coherency:

- **Readability**: assessing the reading level, whether subjectively (using your own judgement) or by using a formula (such as Flesch-Kincaid). Readability formulas analyze word length, sentence structure and length, paragraph size, document length, and so on to provide an overall readability score—an approximation of how easy the text is to read.

- **Consistency**: checking that the same phrases are used to describe identical steps or actions (particularly important when editing for translation) to ensure that key terms are consistently spelled and capitalized, technical jargon is defined, and writing principles are consistently applied.

- **Coherency**: ensuring that the text flows logically and smoothly. This includes checking for clear transitions between paragraphs and sections, maintaining a consistent tone and style throughout, and verifying that all parts of the document support the overall message and purpose.

When to Use: Toward the end of the writing process but before the review and incorporation of feedback. This stage focuses on language, spelling, and grammar, which are elements best refined once the content's structure and technical accuracy are established.

 Note

Readability Formulas

Readability formulas can be misleading. If you're documenting subject matter that uses long and multisyllabic (but familiar or easily understood) words, such as "telecommunications," the readability score may appear poor even though its audience might be fine with it. Readability formulas usually prefer short words and can give a false impression that fragmentary sentences are easy to understand.

 Note

Editing for Translation

Another aspect of editing involves preparing documents for translation, if they are to be translated. Documents destined for translation before publication are edited with particular rigor that focuses on simplification. This approach is designed to make the translation process less time-consuming and expensive (fewer words are more cost-effective) and to reduce the likelihood of terminology causing problems for the translators. For more on this, see Part 9: Translate.

19.2.4 Compliance Checking

Compliance checking involves verifying that content has been written in accordance with any legal, intellectual property, or regulatory and industry standards. These standards may proscribe strict requirements for anything from the correct spelling of another organization's trademarks to the use of a certain shape and sized warning symbol. Highly regulated industries, such as aerospace and medical, have detailed requirements for elements to be included or displayed in a certain way in technical documentation. These requirements vary from region to region. See the Editing Checklist Template for more information about specific items to look for.

When to Use: Toward the end of the review process but before approval. After confirming the document is well-written and technically accurate, it's important to ensure it meets legal and regulatory standards.

19.2.5 Proofreading

Proofreading checks the mechanical aspects of the document without significantly altering the wording. This includes looking at things like spelling and grammar, as well as the consistency of terms and acronyms.

In print publication, this involves an additional step: checking the integrity of the proof (the final version of a document that will be used for printing or publication). In this stage, the proof is compared against the approved final draft to detect any missing pages or sections and to ensure errors haven't been introduced. This is particularly important if the proof has been laid out by a graphic designer. It's easy for text to be copied and pasted into the wrong section or for minor typos to be accidentally introduced.

Proofreading is usually performed as the final step before a significant milestone, such as sending out a document for review or publication.

When to Use: As the final check before approval and publication. This step focuses on the mechanical aspects and the integrity of the document, which should be the last line of defense against errors.

19.2.6 Layout and Format Checking

Checking layout and formatting is often lumped together with proofreading, but it's important to call it out as a separate category because it can be quite involved in technical writing. It's the final step before publication, following proofreading.

In this step, the goal is to ensure that the correct template has been applied; that the styles in the template have been correctly applied to the appropriate elements in the text; that navigational elements such as hyperlinks and tables of contents are functioning properly; and that the appropriate branding elements are in place, for example, the correct version and placement of the organization's logo.

This step also involves checking for any tool-specific issues that arise with particular software packages, such as updating dynamic fields in Microsoft Word. These checks vary depending on the tool used and the final format of the document, such as whether it will be viewed on mobile or desktop platforms. This step can be quite time-consuming, but it's essential to get it right. If users can't view your documents due to a formatting issue, all your hard work may come to naught.

When to Use: The last step before publication. This stage is critical for ensuring that the document is not only error-free but also presented in an accessible and professional manner that aligns with the intended guidelines.

19.3 [Theory] Editing Tools and Inputs

During the editing process, writers rely on several key tools and a variety of inputs. This section explains what these tools and inputs are, and how to use them effectively in your editing.

19.3.1 Editing Tools

The table below explains the key tools used during editing.

Table 28: Editing Tools for Technical Writers

Resource	Description
Editing Checklist	A list of common checks that writers should undertake when editing and proofreading technical documents. Its use ensures documents undergo a rigorous and consistent quality-assurance process prior to publication. Our Editing Checklist Template describes checks applicable to each level of editing. You can customize it to suit your own projects.
Editing Sheet	A list of editorial notes that helps maintain consistency in style and terminology within and across documents, saving you from having to revalidate terms every time you encounter them. This list captures any style decisions made during the editing process. Items commonly found in an Editing Sheet include correctly spelled, punctuated, and capitalized acronyms, technical jargon, product names, user interface elements, trademarks, and role names. Our Editing Sheet Template can be customized and used as the basis for your project. Also known as an *editing stylesheet* within traditional publishing, we avoid this term here due to the potential for confusion with visual stylesheets, which are discussed in Chapter 14: Design Stylesheet.

19.3.2 Editing Inputs

During the editing process, you'll need to keep a number of resources close at hand. Here are some of the typical resources required during editing:

Table 29: Editing Inputs

Resource	Description
Style Guide	Guidelines on grammar, punctuation, layout, formatting, structure, and other stylistic aspects of writing. Guidance varies from country to country, so make sure you're using the correct style guide for your primary audience. An example of a commonly used style guide is the venerable *Chicago Manual of Style*, published by the University of Chicago Press, widely used in the United States.
House Style Manual	An organization-specific style guide. A house style manual contains guidelines for writing in the organization's style, including dos and don'ts, instructions for referring to the organization's trademarks and products, standard legal boilerplate, and the appropriate tone. Although house style manuals usually adhere to general style guides, they often include more specific writing guidelines to address particular organizational preferences.
Technical Writing Style Guide	A publicly available style guide for technical writers referencing or documenting an organization's products. Large technology companies such as Google, Microsoft, and Apple have style guides tailored to their own products. These guides specify how to refer to user interface elements in their products, among other details. They're designed to foster consistency and alignment with an organization's brand. A well-known example is the *Microsoft Writing Style Guide*.

19.3 [Theory] Editing Tools and Inputs

Resource	Description
Industry-Specific Style Guide	A guide that contains rules for a particular industry, such as medicine, journalism, law, or academia. These may be globally or regionally accepted guidelines, depending on the industry. Not all industries have their own style guides, but it's something you need to verify if you're new to the industry. One example is the _Publication Manual of the American Psychological Association (APA Style Guide)_. This style guide is widely used in the fields of psychology, education, and other social sciences.
Brand Guidelines	An organization-specific guide focusing on the consistent application of the organization's brand across documents, usually with an emphasis on visual design. The brand guidelines typically mandate the use of organizational templates and explain the dos and don'ts for logos, colors, fonts, and so on. They commonly also cover writing aspects, blurring the line between house style and brand guidelines. Brand guidelines usually originate in the marketing department. When writing online content, such as help articles, features like fonts, colors, copyright messages, trademarks, and logos need to closely follow brand guidelines.
Dictionary	A dictionary is essential for checking spelling. In most cases, the spellchecker on your authoring tool will suffice. Ensure the language is set to the appropriate location, and be aware of the differences between regions. For example, British English differs from American English. Read more about writing for your geographic audience in Part 9: Translate.

Resource	Description
Access to the Internet	You'll need access to the internet to check the correct capitalization and spelling of company names and trademarks, and to perform basic factchecking.

 Note

What to Do If the Guidelines Conflict

Style guides are hierarchical, proceeding from the specific to the general. This means when you're editing, if there's an ambiguity or conflict, always prefer the more specific style guide, such as a house style manual or brand guidelines. The more general guidelines, such as the _Chicago Manual of Style_, act as a catchall for rules not included in the more specific ones.

19.4 [Practice] Edit Draft

The steps below explain the process of editing a draft document.

19.4.1 Step 1: Determine the Level of Editing Needed

As you develop and mature your draft, you'll edit your document multiple times, making numerous passes to refine and polish the content. During this process, you should find yourself moving from the more substantial levels of editing on the left of the Levels of Editing diagram to the finessing stages on the right.

19.4 [Practice] Edit Draft

The number of editing cycles you undertake depends on the level of rigor required for the document. Some documents naturally require a higher degree of polish, such as those that will be commercially printed, as well as any documents visible to your organization's clients or external stakeholders. Documents intended for use by internal teams may not demand the same level of rigor. Nevertheless, remember that every document you write is a reflection of your skills as a technical writer.

Remember, editing is an iterative process. Even when your draft is mature, you may find yourself making unanticipated updates for example, integrating a significant chunk of new material due to a new feature that's been added at a late stage. In this event, you'll find yourself moving back to the left-hand column in the Levels of Editing. You may need to give the new content a substantive edit, validate it for technical accuracy, and then structurally review the entire document to ensure the new content integrates well.

 Insight

Maximizing Review Efficiency through Proofreading

Remember, a well-edited document respects the time your colleagues set aside for document review. To build lasting professional relationships, you should do everything you can to make good use of that time. A poorly proofed document becomes a distraction, as your review team members will stumble over typos, ambiguous wording, and other issues, rather than focusing on the accuracy of the content.

19.4.2 Step 2: Apply Inputs

While you're editing, keep your inputs handy, especially your style guide. These will guide you in applying style, punctuation, grammar, and tone during editing. When applying the rules, start with the most specific and proceed to the most general. A rough order of precedence is: Brand Guidelines > House Style Manual > Industry-Specific Style Guide > General Style Guide. (The ">" symbol indicates "takes precedence over.")

19.4.3 Step 3: Utilize Tools

As you edit, you should draw on your editing tools: the Editing Checklist and Editing Sheet. Both have different, yet equally important, roles:

- **Editing Checklist:** Use the Editing Checklist Template to remind yourself of the types of checks to perform at each editing level. Customize it for your project. Over time, with practice, these checks will become second nature, and you'll likely not need to refer to it as often. However, it's a useful reminder to ensure nothing is missed, especially when producing multiple smaller documents or topics.

- **Editing Sheet:** As you edit your document or set of documents, use the Editing Sheet Template to record any editing notes you've made during the process. This can include correct spellings of trademarks, product names, user interface elements, role names, technical terms, and so on—anything you've invested time and effort in researching and validating. Maintaining a list of these elements ensures consistency within and across documents and saves you from repeatedly validating the same details. It can also become a valuable resource to be shared with the writing team.

19.4.4 Step 4: Review Draft

Finally, when you're satisfied that your document has been thoroughly edited and polished, it's ready for review by subject matter experts, and perhaps your peers as well. This ensures that it is technically accurate and meets expectations. We discuss the nuances of review in the following chapter: Chapter 20: Review Draft.

Part 8 Review

Ensuring accuracy and clarity of technical documents through expert reviews and validation.

Chapter 20 Review Draft

Lead Writer: Felicity Brand, Kieran Morgan | **Peer Reviewer/s**: Steve Moss

This chapter focuses on the review phase in technical writing. It highlights the importance of peer, subject matter expert, and approver reviews in ensuring document quality. It covers core concepts such as review etiquette, including techniques for diplomatically giving and receiving feedback. The chapter also discusses practical steps for a successful review process and emphasizes the need for technical writers to skillfully manage reviews while maintaining strong working relationships with colleagues.

 Who Should Read This

- Aspiring Technical Writers
- Beginner Technical Writers
- Cross-Domain Professionals

CONTENTS

20.1 Introduction 351
20.2 [Theory] Write-Review-Update Process 352
20.3 [Theory] Types of Review 353
20.4 [Theory] Review Etiquette 355
20.5 [Practice] Review Draft 360

Chapter 20 Review Draft

PROCESS

Inputs	Review > Review Draft		Outputs
• Validated and Tested Information	**Theory**	• Write-Review-Update Process • Types of Review • Review Etiquette	• Request for Review • Reviewed Draft
	Practice	• Define Review Team • Validate and Test • Request Review • Chase Up Outstanding Tasks • Collate Feedback • Update Draft • Close Out Comments • Respond to Reviewers	
	Examples	• Sample Request for Review	
	Templates	• Review Log Template	

20.1 Introduction

Imagine a writer. You might picture a solitary figure hunched over a lamplit desk, pouring their genius into a Great Work. This romanticized image, while traditional, misses the mark when it comes to technical writing. Technical writing is a highly collaborative process involving inputs from numerous people at every stage. In our world, writing is far from a solo pursuit—it's a team sport!

Nowhere is this more true than in the review phase. It's in this phase that the quality of technical content is ensured through review by numerous collaborators, plus a process of rigorous validation and testing. In this process, the writer acts as a conductor orchestrating the end-to-end process, chasing up recalcitrant reviewers, updating and editing documents, and managing progress and timelines along the way. Experienced writers often juggle numerous drafts at different stages of maturity simultaneously, masterfully guiding them through to completion without compromising on quality.

In this chapter, we explain the concepts behind a successful document review, from the nuts and bolts of the different types of reviews—peer review, subject matter expert review, and approval—to review etiquette, which is the art of giving and receiving feedback. This is critical to building good relationships with your colleagues. As we transition to Chapter 21: Validate and Test Information, we'll build on these concepts, focusing on validation and testing methods. These are crucial steps to ensure your documents aren't just well-written but are also accurate and user-centric.

By the end of this chapter, you'll be equipped with practical steps for producing high-quality, thoroughly reviewed documentation. Balancing quality with timeliness can be tricky, but with the right approach, it's achievable. So, let's get started on this journey to mastering the art of review in technical writing.

20.2 [Theory] Write-Review-Update Process

In our discussion of the write-review-update cycle in Chapter 17: Write Draft, we explored the iterative process of drafting documentation, reviewing it for accuracy, and ensuring its readiness for publication.

The diagram below illustrates a generic write-review-update process for a typical document. It's a simplified view of the overall write-review-update process, omitting the loop-backs that can occur at any stage. Use this diagram to frame your understanding of the interplay of responsibilities among the writer, reviewer, and approver as you progress through the chapters in Part 7: Edit and Part 8: Review.

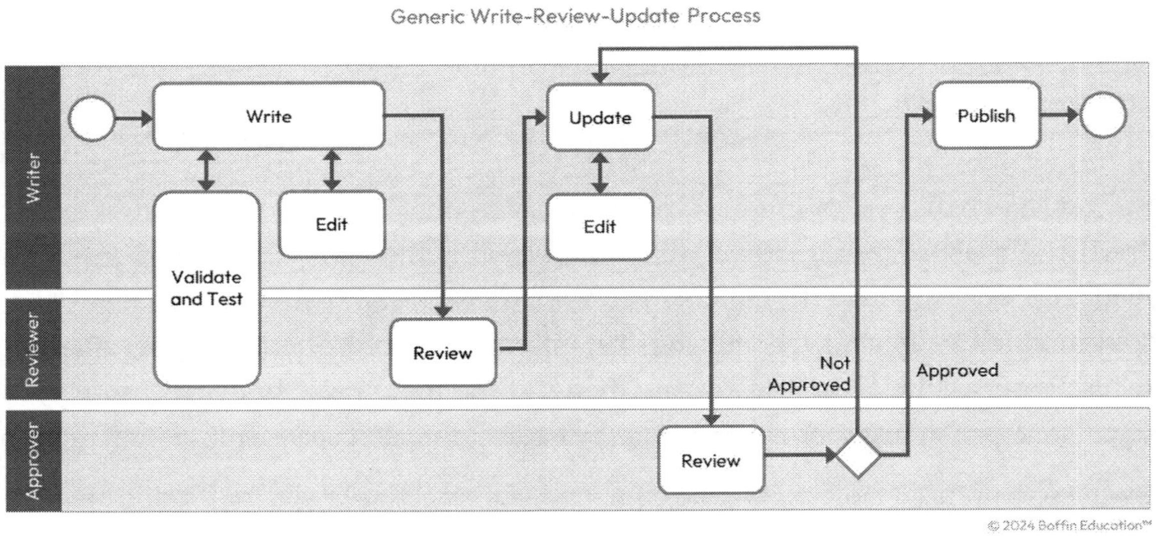

Figure 54 Generic Write-Review-Update Process

20.3 [Theory] Types of Review

In the sample review process above, there are two swim lanes: one for "reviewer" and one for "approver." In a typical document review, these broad-brush swim lanes break down into several unique responsibilities: peer review, subject matter expert review, and approver or document owner review. In this section, we'll explain the nuances of each.

20.3.1 Peer Review

Peer review is a critique of your work by a colleague within your organization—i.e., another technical writer. Peer reviews add significant value: writers can learn from their more experienced or knowledgeable colleagues, and the process helps to converge writing styles among a team of writers.

Rather than a subject matter expert review, which addresses accuracy, usability, and completeness, a peer review analyzes your work through the lens of best practice of your craft and within the context of your organization.

Key Aspects of Peer Review:

- **Document Structure**: Is the document well-structured and logically coherent?
- **Appropriate Wording**: Is the wording appropriate for the context and audience? Does it flow well?
- **Correct Format**: Have you used the correct document template or format?
- **Style Guide Compliance**: Have you applied the in-house style guide?
- **Terminology Consistency**: Have you followed the appropriate conventions for terminology, such as product names and job titles?

The output of a typical peer review is a marked-up document with comments identifying problem areas and offering suggestions for improvement.

> **Tip**
>
> **How to Pick a Good Peer Reviewer**
>
> If you have the liberty of nominating a peer reviewer, select one whose work you know to be of high quality and ideally someone who has been at the organization for a while. It's even better if they're familiar with the product or process you're documenting. Experienced writers will bring valuable insights that will improve your document. If they're unfamiliar with your project, make sure you provide them with enough context so that they can conduct a meaningful review.

20.3.2 Subject Matter Expert Review

The primary purpose of subject matter expert review is to validate your document from a technical perspective, review it for completeness, and determine whether it's suitable for the intended audience. It's not for pointing out typos, or errors in formatting and grammar. These should have already been addressed during the writing and editing stages.

You should also have made every effort to validate and test the documentation yourself before sending it for review. This maximizes the use of your subject matter experts' time by weeding out the most obvious inaccuracies. See Chapter 21: Validate and Test Information for more information.

Key Aspects of Subject Matter Expert Review:

- **Technical Accuracy**: Does the document accurately reflect the technical nuances and complexities of the subject?

- **Incorporation of Updates**: Have any recent updates or changes to product features or technical specifications been appropriately incorporated?

- **Content Accessibility**: Is the technical content presented in a way that is accessible and understandable to the intended audience, without oversimplifying the core concepts?

- **Clarification of Ambiguities**: Are there any technical inaccuracies or ambiguities that need to be clarified or corrected?

- **Industry Standards Compliance**: How well does the document align with industry standards and best practices?

- **Consistency of Terminology**: Are all technical terms used correctly and consistently throughout the document?

20.3.3 Approver/Document Owner Review

The final stage in the review process often involves the approver or document owner—typically a manager, project lead, or department head. This review is vital as it ensures that the document aligns with the overall goals, policies, and standards of the organization. The approver's role is to provide the final seal of approval, confirming that the document is ready for publication or distribution.

Key Aspects of Approver/Document Owner Review:

- **Alignment with Objectives**: Does the document support and align with the broader goals and strategies of the team, department, or organization?

- **Compliance with Policies and Standards**: Is the document in compliance with organizational, legal, and industry standards?

- **Overall Quality and Effectiveness**: Does the document achieve its intended purpose effectively? Is it clear, concise, and useful to its target audience?

20.4 [Theory] Review Etiquette

An essential element of a successful review is knowing how to give and receive feedback diplomatically and graciously. In this section, we explain the ins and outs of good review etiquette. This section goes beyond the immediate requirement for document quality and focuses on establishing a long-term relationship of trust and mutual respect with your colleagues.

> **Tip**
>
> **Be Respectful of Your Reviewers**
>
> Reviews can take a lot of time out of someone's busy schedule, so make the process as easy as possible and don't make unreasonable demands. Consider making large review tasks smaller by breaking big documents into chunks (for example, on a chapter-by-chapter basis) that can be reviewed independently. Build relationships with your reviewers by remembering common courtesies such as thanking them for their time.
>
> If you're not planning on making a change someone has suggested, tell them about it. They need to know that you're taking them seriously and respect their contribution. If you don't respect their feelings, they won't put in the effort the next time you ask.

20.4.1 Receiving Feedback

It's important for you, as the writer, to be prepared to take a step back from your work during review. When you receive feedback, take a deep breath and try not to take it too personally. Although it may sting to receive a critical review, once you recover from the blow to your pride and get down to the business of revision, you'll be able to use that feedback to make your document better. Even bestselling authors heed advice from their editors and consider them invaluable partners in the writing process.

Here are some examples of effective and ineffective responses to feedback.

Table 30: Examples of Effective and Ineffective Reactions to Review Feedback

Ineffective	Effective
- **Ignoring Feedback**: Dismissing feedback without consideration, for example, out of a desire to meet deadlines. This approach prioritizes timeliness over quality—almost always a mistake in technical documentation. - **Defensive Response**: Taking critical feedback as a negative reflection on your overall skill set rather than your document's accuracy and quality. The best writers understand that everyone can benefit from a different perspective.	- **Objective Analysis**: Evaluating feedback based on its merit, not on the tone or the person giving it. Accepting feedback as a reflection on the document, not a general comment on your skill set. - **Collaborative Clarification**: Engaging in respectful dialogue with the reviewer to understand their perspective and clarify any misunderstandings. Gently guiding them toward the best way to give you value-adding feedback.

Here are some techniques to help ensure that you make the most of review feedback:

- **Maintain Perspective**: Recognize that feedback is aimed at improving the document, not critiquing you personally. A critical review is an opportunity for growth, not a reflection of your competence.

- **Evaluate with Openness**: Consider each piece of feedback objectively. Even if the delivery is less than diplomatic, the underlying point might be valid.

- **Verify Technical Details**: If feedback involves technical changes, verify the accuracy with relevant sources or consult with another subject matter expert to ensure consensus.

- **Engage in Dialogue**: If you're unsure about a comment or feel it's based on a misunderstanding, engage in a conversation for clarification. This can lead to a better understanding of the reviewer's perspective.

- **Document Decisions**: Keep a record of feedback and your responses, such as comments in your authoring system or a *Review Log Template*. This not only helps in tracking changes but also in explaining your decisions to the review team.

- **Acknowledge and Thank**: Always thank reviewers for their time and input, regardless of whether you implement all their suggestions. A polite acknowledgment of their effort fosters good working relationships.

20.4.2 Giving Feedback

Reviewing another's work can add tremendous value to a document, but it can strain the relationship if the review isn't performed in a sensitive manner. Even if a reviewer doesn't intend to—in fact, even if all their review feedback is completely accurate—they may cause offense simply due to the way their feedback is phrased. Aspire to be both a gracious recipient of feedback and a diplomatic giver of it and your relationships with your peers and colleagues will thrive.

Here are some examples of poorly and well-phrased feedback.

Table 31: Examples of Poorly and Well-Phrased Review Feedback

Poorly Phrased Feedback	Well-Phrased Feedback
"Your introduction is confusing and lacks clarity. You need to rewrite it."	"The introduction has some great ideas, but it might benefit from a bit more structure to guide the reader through the concepts more clearly. Perhaps an outline of the main points could help?"
"This section is too technical. Simplify it immediately."	"I appreciate the technical depth in this section, but it may be a bit complex for beginners. Could we consider adding some introductory explanations or simplifying some of the technical terms for accessibility?"

So, what went wrong? Both sets of comments are essentially conveying the same points, but the poorly phrased comments are written in a hurtful way that's not constructive. They also don't offer concrete solutions, just negative feedback.

To avoid the situation above, here are some general rules on giving a good peer review:

- **Do It Justice**: Commit to reading and comprehensively understanding the document. Avoid merely skimming, as it might lead to superficial or inaccurate feedback.

- **Note First Impressions**: Keep a record of your initial reactions as you review the document. These first impressions are valuable and can provide a strong foundation for your formal review. Technical writers are often the first users of a product, so your initial reactions will likely mirror those of the target audience.

- **Guide with Suggestions**: Rather than issuing commands, guide the writer toward better practices with suggestions and questions, as illustrated in the examples above. This approach is effective whether you're a colleague or a manager, as it nurtures the writer's sense of autonomy, responsibility, and pride in their work.

- **Acknowledge When You Don't Know the Solution**: You're not always going to have the answer, but identifying when something isn't quite right can also be useful. For example, you could say, "I keep stumbling over this part but can't think of any suggestions for how to fix it."

- **Be Specific**: Ensure your comments are clear and directly related to specific sections of the document. This is particularly important when responding via email, as opposed to inline comments within a document. In emails, it may not be immediately obvious which section you are referring to.

- **Balance Negative with Positive**: Always find aspects of the document to praise. A reviewer's role is to be constructive, aiming to enhance the document. Acknowledging what works well is as important as identifying areas for improvement.

- **Recognize You Might Be Wrong**: As a reviewer, present your feedback as your personal opinion rather than as a universal truth. Who knows, you might be wrong! For instance, a house style rule or technical detail you're familiar with could have recently been updated without your knowledge.
- **Follow Up and Engage**: Encourage a dialogue with the writer after delivering your review. Offer to discuss any points that might need clarification and be open to understanding their perspective. This follow-up strengthens the relationship and ensures that feedback is well received and understood.

20.5 [Practice] Review Draft

Now that you understand the importance of the different types of review, it's time to get your documents reviewed.

20.5.1 Step 1: Define Review Team

If you haven't already done so, you'll need to define your reviewers and their responsibilities in a review matrix. Receiving a request for review shouldn't come as a surprise! We discuss this planning step in detail in Chapter 12: Define Review Team, and provide a Sample Review Matrix to capture all the necessary information to prepare for a successful review. This helps you plan out your review in advance, ensuring it doesn't delay your publication date.

 Tip

Be Choosy about Your Experts

It's essential to have the right people review your content. A poorly informed subject matter "expert" can cause more problems than they solve. Make sure you're reaching out to experts with a true depth of knowledge in the subject you're documenting—and ensure they're fully engaged in the review process.

20.5.2 Step 2: Validate and Test

Before you send your documents for review, make sure they're validated and tested. This will catch errors and inaccuracies in advance of expert review, which will save time for your reviewers and reduce the number of review cycles required. It will also reflect well on your craftsmanship if you send out a well-researched document. In Chapter 21: Validate and Test Information, we showcase different methods for doing so. These methods vary in complexity and cost, and their application should be strategically timed—some before the subject matter expert review, some after, and others at any stage.

20.5.3 Step 3: Request Review

When it comes time for review, many folks fire off a hastily worded request for review email, attach their documents, and hope for the best. However, you can get much better results by acting as a guide for your experts. Prompt them into giving you good, value-adding feedback by explaining the aim, purpose, and mechanics of the review. This will help them focus on the areas that you want them to—and steer clear of those that you don't want feedback on, such as grammar and punctuation.

When sending documents for review, frame the accompanying note with care to ensure the review team understands what's required. Include the following:

- **Context**: It shouldn't be a surprise to receive your request. However, if your experts are the kind of people who are frequently inundated with review requests, jog their memory with a concise précis of your document's purpose.
- **Timeframe**: Include a clear timeframe by which a response is required. If you're working in an Agile environment, earmark a particular sprint you'd like to receive the feedback in.
- **Feedback Mechanism**: Specify how you'd like to receive their feedback. Should they use comments and track changes in an attached document? Or the review functionality of your organization's authoring tool? Do they need to respond to a workflow request in a content management system?

- **Review Responsibilities**: What are their respective responsibilities? Guide them to focus on their area of expertise so they can add the most value with the minimum wasted effort. This should be in alignment with the review matrix in your Documentation Plan Template.

- **Option to Approve**: If they're satisfied with the document as-is, how do they indicate their approval?

- **Thank You**: Thank them in advance for their time to show that you value their input and the time investment they make in reviewing your content.

See Sample Request for Review for an example of a well-crafted request for review. Use it as a template or for inspiration when you're crafting your own request.

 Insight

Comments vs. Markup

Some writers prefer to request comments from subject matter experts instead of marked-up changes. Comments provide scope for detailed explanation, which you don't get with markup. This lack of context can leave writers with more work as they chase up the intention behind a marked-up change. Also, track changes can quickly become unworkable if revisions from several reviewers need to be integrated back into a single document.

 **Note**

Doc Reviews in a Docs-as-Code Environment

If you're working in a docs-as-code environment, your review mechanism with developers may be through a process known as pull requests (PRs) or merge requests (MRs). In this setup, documentation resides inside (or alongside) the code, typically in the form of simple text files. Developers use PRs/MRs to review and provide feedback on your content. They may directly edit the content or offer feedback through comments in the code.

20.5.4 Step 4: Chase Up Outstanding Review Tasks

A seldom-acknowledged aspect of being a great technical writer is learning how to be a persistent chaser. The most productive technical writers are almost always following up on review tasks, as they'll often have multiple documents under review simultaneously. This may seem tedious, but it's all part of the job description. Consider it an exercise in the art of diplomacy. Learn to be polite yet persistent. It's a balance between ensuring timely publication and technical accuracy without causing undue friction among colleagues.

Here are some methods you can use to chase up outstanding review tasks, in sequential order:

1. First, see if reviewers respond within the timeframe. There's no need to badger anyone before the timeframe has elapsed, unless you receive new information and the need becomes urgent.

2. Once the original timeframe for review has elapsed, immediately send a reminder note or email as a nudge.

3. If you haven't received a response to your note within a few days, follow up with a phone call, a message on an internal messaging system, or a text message to their phone number. This is more personal—and harder to ignore—than an email. Be polite but clear—explain that the request is now urgent and holding up publication.

4. If you still haven't received a response after trying all these methods, request assistance from your line manager or a project manager. Note the times and dates you tried the methods above, showing them the emails if necessary. They can then use their authority and relationship with the expert, or the expert's manager, to apply the appropriate pressure.

Tip

Fallback Methods When You Just Can't Get a Response

What if you've tried everything and you still can't get reviewers to respond? Unfortunately, this can happen. Here are some fallback methods you can use if all other techniques have been to no avail:

- **Review Workshop**: This is a technique for dealing with time-poor subject matter experts. It involves scheduling a short (typically one- or two-hour) workshop with a group of experts. As the facilitator, you walk the experts through the documentation on a shared screen. Don't hesitate to ask questions or challenge parts of the document that seem problematic. Keep a record of the discussion and decisions made by using something like comments or sticky notes. This technique is excellent for yielding first impressions and identifying obvious errors, but don't expect an in-depth examination. It works best for shorter documents, such as processes and brief procedures.

- **Endorsement by Default**: When you send out a review request for low-risk content, consider including a "default endorsement" clause, for example: "If no feedback is received by [date], I will assume your endorsement." Such a statement clarifies what will happen if a reviewer doesn't provide feedback. This approach can be helpful where you've thoroughly tested and validated content already and the review is more of a formality. However, don't use this technique for high-risk content, like safety procedures, or high-impact content, such as user guides.

20.5.5 Step 5: Collate Feedback

Now that you've completed a thorough review and followed up on outstanding review tasks from your reviewers, it's time to collate their feedback. If you use a system that aggregates all feedback in one place, excellent! You'll save a lot of time, as the system has already done this for you. This functionality is common in many content management systems and authoring tools.

However, you may find that—for various reasons—you receive feedback in diverse forms. This is where a Review Log Template can be useful. This spreadsheet allows you to keep track of comments received, associate them with a specific version of the document, and monitor the actions you took to address them.

The Review Log is a valuable tool for consolidating feedback, especially if you receive review comments from different channels—such as email, sticky notes, written markup on hard copies, or verbal feedback noted down during the morning stand-up. It's also helpful when managing a highly controlled document where you need to detail your response to every comment.

Tip
Strike a Balance between Organization and Efficiency
With our passion for attention to detail as technical writers, it's easy to get carried away with administrivia. However, there's usually no need to log absolutely everything in your Review Log. Aim to strike a balance between the benefits of having a single source of truth for review feedback and the extra time it takes to consolidate and process all the commentary you receive.

What Does That Mean?

Feedback

Any comments or markup received from a reviewer in a document review.

Markup

Symbols such as strikethrough and "red-line" that indicate changes to a document, such as additions, changes, and deletions. These may be written on a hard-copy printout or applied via the track changes functionality in an authoring tool or word processor.

Comments

Written notes about any aspect of a document. These may be specific to a section or paragraph within the document, or general commentary about the document's overall effectiveness. These may be emailed as a summary, affixed to a hard copy with sticky notes, or applied via the comments functionality in an authoring tool or word processor.

20.5.6 Step 6: Update Draft

Now that you've consolidated all the review comments, it's time to update your draft document. Remember, review is a two-way street. It's up to you, as the writer, to decide whether or not to accept the suggestions from the reviewers. Don't take every comment as gospel; carefully weigh each comment on its merits before acting. Consider whether it adds to or subtracts from the overall usability of the document, and carefully validate any changes to technical details before implementing them.

20.5.7 Step 7: Close Out Comments

As you work through the feedback, note your response to each comment received and mark them as resolved. This way, both you and the reviewers know you've actioned them—even if it's just giving a thumbs-up or marking it as complete. This approach helps you keep track of everything. If you need to write a response—for example, your rationale if you've decided not to action a comment—make your comments concise yet diplomatic, and ensure they're archived somewhere visible to reviewers. Whether in a shared document or attached and sent via email, it will give reviewers the option to respond with further clarifications if necessary.

20.5.8 Step 8: Respond to Reviewers

Finally, respond to the reviewers with a brief thank you note. This can be as formal as an email or a brief note on an internal messaging system. Think of this as a gesture of your appreciation for the time and effort they've put into reviewing your document. It also involves them more deeply in an ongoing conversation about your document's accuracy and usability. Although some folks won't mind either way, it's easy for people to feel offended if they think you've ignored their feedback, which will make them much less likely to contribute meaningfully to a review in the future.

Chapter 21 Validate and Test Information

Lead Writer: Kieran Morgan | **Peer Reviewer/s**: Felicity Brand, Steve Moss

This chapter emphasizes the critical role of validation and testing in technical documentation. It details various methods like factchecking, testing, user acceptance testing (UAT), usability testing, and informal usability testing. These are all important steps for ensuring technical accuracy and effectiveness in documentation. The chapter aims to equip technical writers with the necessary skills to create not only precise but also practical and user-centered documentation.

 Who Should Read This

- Aspiring Technical Writers
- Beginner Technical Writers
- Cross-Domain Professionals

CONTENTS

21.1 Introduction	371
21.2 [Theory] Factchecking	372
21.3 [Theory] Testing	372
21.4 [Theory] User Acceptance Testing (UAT)	374
21.5 [Theory] Usability Testing	375
21.6 [Theory] Informal Usability Testing	377

Chapter 21 Validate and Test Information

PROCESS

Inputs	Review > Validate and Test Information		Outputs
• First Draft	**Theory**	• Factchecking • Testing • User Acceptance Testing (UAT) • Usability Testing • Informal Usability Testing	• Validated and Tested Information
	Practice	• See Chapter 20: Review Draft.	
	Examples	• N/A	
	Templates	• Review Log Template	

21.1 Introduction

Let's say you're a scientist on a voyage of discovery within your field of expertise. Maybe you've dreamed up a new molecule that you think is going to rid the world of some terrible disease. You've done the research, read the academic papers, and worked through your equations. All the modeling shows you're on the right track.

But you haven't yet tested your theory in the real world. Is your theory valid? Not yet! That's not how the scientific method works. As a scientist, you can't claim your theory is effective until it's been rigorously tested under experimental conditions. Similarly, in technical documentation validation and testing are critical phases where the "theory" of your documentation meets the realities of the product and the user experience. It's not just about technical accuracy; it's also about effectiveness. Does your documentation enable your audience to successfully complete tasks?

In Chapter 19: Edit Drafts, we touched on these concepts as vital levels of editing in technical writing. Validation and testing are not stand-alone processes but integral elements of the comprehensive review process outlined in Chapter 20: Review Draft. The sequencing of these steps—when they're done in practice—is illustrated in Chapter 20: Review Draft: Write-Review-Update Process.

In this chapter, we explore in depth the methods of validation and testing for technical documentation. These methods vary in complexity and cost, and their application should be strategically timed. Some come before the subject matter expert review, some after, while others can be scheduled at any stage. Your judgment in choosing the right method for your documentation will significantly influence its accuracy and user-friendliness.

Through this chapter, we aim to equip you with the knowledge necessary to ensure that your documentation is not only precise but also practical. It should enhance the user experience and contribute to the success of the product it supports.

> **Insight**
>
> **Add Value by Identifying Issues and Bugs**
>
> Technical writers are often the first users of a product. While you're testing your documentation, you may discover usability issues—or bugs—in the product you're documenting. By providing this feedback to the developers—e.g., by logging it in their development tracking software—you can add value by improving the user experience before the product is released. Being a diligent and proactive member of the development team can be a career-enhancing move for a technical writer.

21.2 [Theory] Factchecking

Factchecking involves verifying technical facts against specifications, data, reference documentation, or by verifying them with key subject matter experts. For instance, you might confirm that a range of values described in a product's user guide matches those in the product specifications. This is a crucial step for technical documentation, and it's something you can easily do yourself by making sure you can validate every technical fact with reference to another document. Use this process to showcase your attention to detail and commitment to accuracy.

When Is It Done?

- Prior to review by subject matter experts, or simultaneously, if requesting validation of technical facts from experts.

21.3 [Theory] Testing

Testing is a process you can do yourself. It's simple: see if you can follow your own instructions to operate the product (hardware or software) that you're documenting. In doing so, you should also check that images or screenshots of the product match those in your document at each step.

21.3 [Theory] Testing

This step is essential for product documentation. If you can't follow your own documentation, how would a customer hope to do so? This process will also help you quickly identify whether your procedures are clearly worded and logically structured. It's an easy and inexpensive way of ensuring high-quality documentation.

If it's not possible for you to test your procedures yourself, arrange a walkthrough with a subject matter expert familiar with the procedure and observe as they follow your steps. This is useful when experts have access to systems and test data that you don't.

As with factchecking, testing is a rigorous and methodical process that draws on your attention to detail. You'll need to step into the shoes of the users, which can be challenging if you've built up detailed product knowledge. For this reason, sometimes additional methods are employed that eliminate your biases and blind spots. These are discussed in the next section.

When Is It Done?

- Prior to review by subject matter experts.

 Insight

Try to Mimic User Behavior When Testing

When testing a procedure, aim to mimic the behavior or use case of the user as closely as possible. Although it's also helpful to test out of context, it's more useful to approach testing from the user's perspective. Challenge yourself to step into their shoes by posing questions: When would a customer attempt a particular procedure? What steps do they need to successfully accomplish beforehand? What would they do when confronted with the next screen? And so on.

>  **Note**
>
> **Check You're Testing the Correct Software Version**
>
> While testing, it's essential to work with the correct version of the product's software or firmware. This is particularly crucial in an Agile development environment, where features may be added or withdrawn at the last minute.

21.4 [Theory] User Acceptance Testing (UAT)

In user acceptance testing, your documentation is tested as if it were software code. In this particular procedure, a testing team develops a number of test cases based on your documentation. They then attempt to execute these cases to see whether the documentation matches the functionality and user interface of the product.

The test cases may simply be a copy-and-paste of procedural text from the documentation, copied wholesale or broken into bite-size chunks. The testing team will then attempt to follow these instructions. If the instructions don't produce the intended result, or if the pictures or text in the documentation don't match the user interface of the product, they'll chalk it up as a failure and pass it back to you for correction. Depending on the available time, the testing team may choose to test only a portion of your documentation, (such as the sections concerning the most high-profile functionality), similar to an audit.

The advantage of this method is that it removes the biases and blind spots you might have when testing your own documentation. It can be done in-house, leveraging your organization's testing infrastructure.

When Is It Done?

- Following review by subject matter experts.

> **What Does That Mean?**
>
> **Test Case**
>
> A scenario describing a set of steps with a defined start and end point, and an expected result. Test teams can use a test case, along with test data that is sufficiently representative of reality, to verify that a subject, such as a software application and its technical documentation, meets a specific requirement.

21.5 [Theory] Usability Testing

Usability testing involves observing users as they attempt to use the product and follow your documentation. This is the gold standard for road-testing documentation. It eliminates any blind spots that you or an in-house team may have due to existing knowledge.

This type of testing is usually the domain of user experience (UX) designers, who test the usability of a product as a holistic package prior to launch, including documentation. Nevertheless, it's worthwhile to understand how the process works, as technical writers should be integral to it. This is why we've included a brief description here. We've based our usability testing content below on Markel and Selber's description in their book *Technical Communication*.[46]

Usability testing can be quite elaborate, with realistic mock-ups of the product and its documentation set up, plus test participants recruited, akin to a scientific study. It can even be conducted in context—that is, at the place where users will be using the documentation. The study may be conducted by an in-house team or by a third party who specializes in usability testing.

The goals of the study and the evaluation guidelines should be defined well in advance. Test participants and observers are identified and recruited, and consent forms are signed. An evaluation form is drawn up so that observers can record any problems discovered during the test. A script and checklist should be drawn up to guide the sequence of events on the day, ensuring consistency among test subjects. On the day of the study, the observer should be neutral, asking open-ended questions, never prompting the user, even if they get stuck. The aim is to understand why the user got stuck and at which point.

After testing has concluded, feedback from the observers is compiled into a report and handed to the product development team. Both the product and its documentation are then updated based on the test's findings.

When Is It Done?

- Following review by subject matter experts.

 Insight

To Test or Not to Test—Do the Benefits Outweigh the Costs?

Usability testing is an elaborate and expensive affair. For this reason, it's rarely used in practice.[47] So, when should it be used? When the importance of the product and its documentation warrants it.

Consider the consequences of users misunderstanding something. Could this lead to thousands of costly helpdesk calls? Is it for your organization's flagship product, where errors could incur significant costs and damage your organization's reputation and brand? Assigning a dollar value to documentation might seem like mission impossible, but it's definitely within the realm of possibility. Consult the Calculate Value chapter in our forthcoming book, *Technical Documentation Management*—available online at https://boffin.education/category/technical-writing/technical-writing-books/—for guidance.

21.6 [Theory] Informal Usability Testing

If formal usability testing isn't an option, you can conduct informal usability testing—with much less cost and effort—by doing it yourself.

To do so, you'll need to engage a proxy user within your organization, or within your social circle, if your organization permits it. This individual should have general demographics or a level of knowledge about the product or process that matches those of the intended audience of your documentation. In this informal process, you don't need to do all the preparation of a formal study. Simply assume the role of an observer, and ask your proxy user to follow the steps of your documentation, noting down any problems they encounter along the way. As you repeat this process, you'll discover many useful insights into the effectiveness of your documentation. Use a *Review Log Template* to capture and action the feedback.

Of course, this method introduces some issues that formal usability testing avoids, such as a smaller pool of testers, the absence of a scientific method, and potential biases as the content's author. Nevertheless, it's a quick and cost-effective way to gather valuable feedback on the effectiveness of your documentation. It's ideal for small teams or those with limited time and budget.

When Is It Done?

- Either before or after review by subject matter experts.

Insight

Having Trouble Following That Procedure? Try Talking It Through.

The talk-through approach is a useful method that helps eliminate bias. In this approach, the technical writer asks a test subject to describe their thoughts aloud as they work through the steps of a procedure. This can help identify any gaps in your instructions or false assumptions you might have made about the product or the user's knowledge. The talk-through method is also useful during formal usability testing.

Part 9 Translate

Mastering translation and localization for global technical communication.

Chapter 22 Translation Theory

Lead Writer: Alison Pickering | **Peer Reviewer/s**: Kieran Morgan | **Expert Reviewer/s**: Stephanie Riches Harries

This chapter explores the concepts of translation and localization in technical writing projects. We emphasize how crucial it is to grasp language subtleties and cultural differences. This understanding is vital for accurately translating technical documents for a worldwide audience. By examining these theories, this chapter equips you with the expertise needed to make sure your documentation is linguistically precise and culturally appropriate. This theoretical foundation is necessary for the practical application discussed in the following chapter, Chapter 23: Translation Practice.

 Who Should Read This

- Technical writers, professionals, and project managers who are responsible for overseeing the translation and localization of technical documentation.

CONTENTS

22.1 Introduction	383
22.2 The Importance of Translation	383
22.3 [Theory] Translation, Localization, Internationalization	385
22.4 [Theory] Writing for Translation	387
22.5 [Theory] Setting Up Images for Translation	389
22.6 [Theory] Scope of Translation Services	390
22.7 [Theory] Roles and Responsibilities in the Translation Process	395

Chapter 22 Translation Theory

PROCESS

Inputs	Translate > Translation Theory		Outputs
• Source Text • Source Images	**Theory**	• Translation, Localization, Internationalization • Writing for Translation • Setting Up Images for Translation • Scope of Translation Services • Roles and Responsibilities in the Translation Process	• Translated and Localized Content
	Practice	• See Chapter 23: Translation Practice.	
	Templates		

22.1 Introduction

This chapter explores the concepts of translation and localization in technical writing. Understanding these is crucial for anyone involved in producing technical documentation that resonates with a global audience. In this chapter, we explore the intricacies of language and culture, and how they influence the translation process. We'll examine the importance of cultural sensitivity and accuracy in translation, as well as how these elements contribute to the internationalization of technical documents.

This theoretical framework sets the stage for practical application discussed in Chapter 23: Translation Practice, where these theories are put into action. Whether you're a technical writer or a project manager, or are involved in any capacity with translating technical content, this chapter will equip you with the essential theoretical knowledge to navigate the complexities of translation and localization.

 Note

In-House and External Translation Providers

Our translation chapters are written from the perspective that your organization will use an external translation service provider. If your organization has an in-house translation team, speak to the translation manager to find out which processes are in place and which tools are used.

22.2 The Importance of Translation

Over the last few decades, we have seen the globalization of markets, making the need for translation of product documentation greater than ever before. Translation and localization provide your business with access to the global marketplace.

Legal requirements vary between industries, and it is your responsibility (or the translation manager's) to learn the legal obligations related to your products or services for your target markets. For example, for technical products that fall under the EU Machinery Directive,[48] user documentation must be delivered in the language of the country where it is being sold.

Good-quality translations may also have a positive impact on the overall customer experience. Where several companies offer similar products at similar prices, the localized product documentation can be seen as a differentiating factor.

If customers do not understand how to use a product due to poorly translated instructions, they will contact customer support more often, which in turn increases costs for the organization. Taking it one step further, actual errors in translated product documentation can lead to serious consequences for manufacturers, distributors, and the entire business, especially if the errors relate to safety issues. Errors in translations can also cause the reader amusement—and not in a good way. Customers will stop reading a poor translation. You've lost your audience, and their opinion of the product can suffer as a result. Therefore, it is essential to use professional technical translation services rather than just getting your local sales company or distribution agent to translate for you.

When translating technical documents, there are certain tips that can help you achieve the highest possible quality of translation. Remember that this is a process where the quality of the final output depends on the quality of the input, so it's important to get that right. The next section highlights the main points to consider when writing for translation.

22.3 [Theory] Translation, Localization, Internationalization

22.3.1 What Is Translation?

Translation is the process of converting the text of a written document from one language into another while maintaining the original message and communication. For the best results, the translator should be a native speaker of the target language and have an excellent command of both languages involved. This is, of course, necessary for any type of translation, but it is especially crucial when instruction manuals are technical and must be precise and clear. Translation is not necessarily a word-for-word conversion, but it must preserve the original meaning and respect syntax and grammar rules.

For translations of specialized content, such as in engineering or manufacturing, the translator must have specific knowledge. Specialist fields often have their own jargon, and it is these specific terms that make translating in specialist fields more complex; therefore, expertise is essential.

The importance of high-quality translations should not be underestimated. Using cheap and fast services usually results in poor-quality translations, which can alienate the target audience and may also result in higher costs in the long run.

22.3.2 What Is Localization?

One of the most common questions asked is whether documents should be translated or localized. They are closely related, so it isn't surprising that sometimes the distinctions between them are unclear.

Localization is just as important to a business as translation. Whereas translation refers to the process of changing text from one language to another and achieving an equivalent meaning, localization takes the process one step further. It focuses on making text both linguistically and culturally accurate to a specific region, allowing your company to cross cultural barriers and really connect with your audience.

Localization encompasses several areas, such as:

- Spelling (e.g., US English = specialized, UK English = specialised)
- Vocabulary (e.g., US English = elevator, UK English = lift)
- Imperial vs. metric measurements (e.g., US English = inches, UK English = centimetres)
- Currency units (e.g., US English = $, UK English = £)
- Date formats (e.g., US English = MM-DD-YYYY, UK English = DD-MM-YYYY)
- Linked URLs that are country- or region-specific
- Country- or region-specific content (often for legal or regulatory requirements that vary)

22.3.3 What Is Internationalization?

Internationalization is the process of generalizing a product, software, or documentation so that it can handle multiple languages and cultural conventions without the need for redesign or rework. Internationalization practices save time and money, as the localization of a product or communication takes twice as long and costs twice as much if not properly internationalized.[49]

Although the technical writer would not normally be involved in creating or approving product and feature names, it may be worth checking with the stakeholders (usually the marketing department) to ask if a new name to be used in a technical document has been vetted with regional offices where Languages Other Than English (LOTE) are spoken. This strategy, when used early in the process, can prevent the painful renaming and need to rework documentation containing the new brand, product, or feature name. Failure to include this step has caused companies serious blunders over the decades because they didn't check outside of their source language bubble.

Example

German luxury car manufacturer Mercedes-Benz decided to introduce its cars to the Chinese market under the shortened name "Bensi." However, this word means "rush to die" in Chinese, which is not the image Mercedes-Benz wanted to promote. The company quickly rebranded to "Benchi," which means "run quickly as if flying."[50]

22.4 [Theory] Writing for Translation

In order to get a reliable translation of your technical documentation, you need to provide high-quality and simple source text and apply the principles below, as well as the general best-practice principles detailed in *Part 6: Write*. This will guarantee a smooth flow of the text, regardless of the language.

- **Use Active Tense and Simple Language**
  - Prefer active tense, such as "Connect the brake," not "The brake must be connected."
  - Keep sentences short and straightforward, splitting longer sentences into shorter ones or bullet points for clarity.
- **Consistency in Terminology**
  - Maintain consistent terminology throughout the text. Avoid using multiple terms for the same concept to prevent confusion.
  - Eliminate ambiguity by using explicit language, ensuring clarity in translation.

- **Avoid Complex Constructions**
  - Steer clear of abbreviations, acronyms, and Latin terms unless they are widely recognized in the industry.
  - If abbreviations or acronyms are necessary, introduce them with their full wording initially, followed by the abbreviation in brackets.
- **Streamline Text for Efficiency**
  - Reuse content wherever possible for uniformity and cost efficiency.
  - Reduce the number of words to lower translation costs and simplify the translator's task.
  - Avoid forward slashes, preferring explicit connectors like "and" or "or."

You should be thinking about translation, localization, and internationalization as you create and format the source content. Having good-quality input will lower your translation costs, reduce turnaround time, and improve the translation quality.

To help you apply these principles, we've prepared a handy checklist of dos and don'ts: Translation Best Practices Checklist for Text and Graphics.

 Note

Some Templates Are for Subscribers Only

Some of the templates in this book are for subscribers only. You'll need to subscribe to our online knowledge base at https://boffin.education/ to access them in editable format. Subscribers to the e-book and paperback versions of this book can use the discount code in Templates to obtain a free one-year subscription and access the full breadth of our technical writing content. If you don't want to subscribe, head over to our website. There you'll find many of the more straightforward templates available for free, and others are presented in a noneditable format as images.

22.5 [Theory] Setting Up Images for Translation

When preparing images for translation in technical documentation, it's important to consider how they will be interpreted and altered in different languages. This involves not only the textual content within the images but also their overall design and layout. Adhering to certain principles can significantly streamline the translation process, reduce costs, and maintain the integrity of the visual message across various languages.

- **Minimizing Text in Graphics**: Design graphics to be as language-neutral as possible. Avoid embedding text in images, as this necessitates creating multiple versions for different languages and increases desktop publishing work. Instead, use callouts linked to a separate legend or key.

- **Editable Text in Graphics**: If text in graphics is necessary, ensure it's easily editable. For example, by placing text in layers in Adobe Illustrator, direct alteration in the graphic file is facilitated, smoothing the translation process without the need to recreate the image.

- **Accommodating Text Expansion**: Allow for text expansion in graphics. Translated text can be longer than its English counterpart, so designing with extra space prevents overcrowding and preserves readability in all language.

Chapter 22 Translation Theory

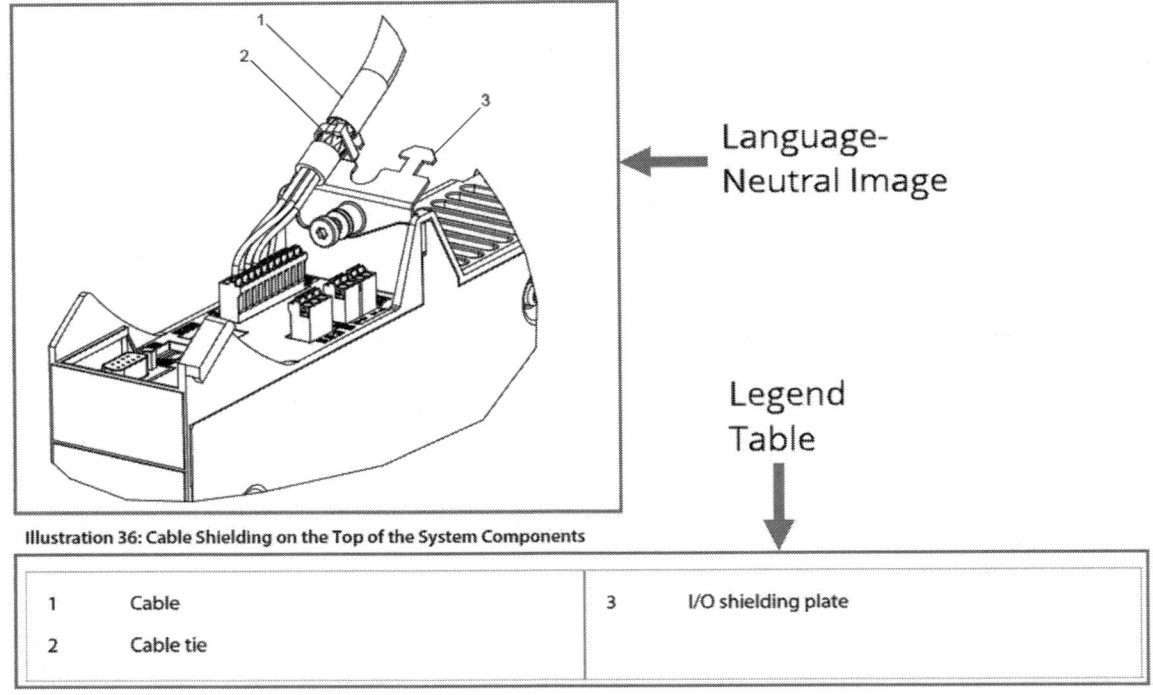

1	Cable		3	I/O shielding plate
2	Cable tie			

Figure 55: Example of a Language-Neutral Image for Translation

For a comprehensive list of dos and don'ts when setting up images for translation, refer to the checklist, *Translation Best Practices Checklist for Text and Graphics*.

22.6 [Theory] Scope of Translation Services

The scope of translation services varies, so it's important to decide what best suits your requirements. It is not generally the responsibility of the technical writer to define the scope of translation services required, but it's good to be aware of the process.

Usually, it is the translation manager/translation responsible who decides on the scope while liaising with internal stakeholders, such as sales and product management, where required. It is vital to be specific about the requirements when requesting a quote to avoid unexpected costs later.

More details can be found in Chapter 23: Translation Practice: Defining the Scope of Translation.

22.6.1 Translation, Editing, and Proofreading

Translation should be done by qualified translators who are native speakers and subject matter experts.

Editing involves reviewing a translation and making changes to improve its quality. This improves the flow and results in a more cohesive translation. It usually focuses on style and content.

Proofreading is the process of reviewing the final draft to verify that translations are accurate and complete, and it primarily focuses on ensuring the translation is grammatically correct (four-eye principle). A proofreader checks the following:

- Overall readability
- Grammatical errors and typos
- Accuracy of the translation compared to the source content
- Style, formatting, and layout are according to the customer brief

If you use a professional translation service provider to translate your content, they will usually also undertake the editing and proofreading process unless otherwise instructed. Therefore, when ordering translations, remember to define the scope.

22.6.2 Machine Translation (MT) and Post-Editing

Machine translation (MT) is a cost-effective and time-saving method; however, it can lack the fluency and specialist knowledge of a human linguist. As its name suggests, it uses machines to translate from one language to another. Although it was initially used as a quick and low-cost method for information translations for internal use, its use in more technical areas and for external documents such as product documentation is increasing thanks to advances in technology and reliability.

While the advantages of MT are cost savings, time, and scalability, there are also some disadvantages you need to consider. In terms of accuracy, machine translation will never match the accuracy and reliability of a human translator's expertise, experience, or industry knowledge. Machine translation doesn't account for local cultures or customs and cannot handle figures of speech, which risks some meaning getting lost in translation. Machine translation is supported by translation memories, so it may be more suitable further down the line when the translation service provider has completed several translations for you, which are then saved in the translation memory.

Post-editing is used following machine translation, and it is a vital step that should not be overlooked. This is the manual human editing and revising of the raw output from a machine translation engine, combining the speed of machine translation with the accuracy and expertise of human translators. During the post-editing step, an experienced linguist will identify and correct words or phrases that the machine translation engine has produced that don't make sense in the context of your text. They may also notice that the text needs to be localized to be acceptable in the target culture.

22.6 [Theory] Scope of Translation Services

 What Does That Mean?

Translation Memory

A database used by translation service providers to store completed translations. It is utilized in subsequent translation tasks to identify and automatically translate repeated text, ensuring consistency and reducing the need to retranslate identical content, thereby saving time and costs. Translation memories can be shared between different service providers, facilitating efficiency when multiple vendors are used.

22.6.3 Validation, In-Country Review, Client Review

Validation, also known as in-country review or client review, is a process to check the correctness of translations for the specified (or target) country or region. Once the translation is completed, it is passed to a designated validator who reviews the translation and suggests improvements. This may be an employee of your local sales company or a product manager who is a native speaker of the target language and also a subject matter expert. You should define primary and backup validators for each country and product, and communicate this to the translation service provider.

Validation can be done on both human and machine translations. Most translation service providers have a software tool for this purpose, and validators should be offered training in using it.

During validation, the validator should be given clear instructions about their task and the deadline. They should be informed about what they should check for and flag, and what they should ignore. Some SMEs are so passionate about their area of work or product that they go beyond what is requested and even start making changes to content that are not in the source text, such as adding sentences or changing technical data. An example of what a validator should check includes:

- Accuracy of the translation

- Correct localization of the text
- Consistency with approved terminology for their language

Regardless of whether the validation stage is handled internally or by the translation service provider, the validators should always be given access to the full set of approved terminology—the terminology database (or termbase)—to support them in their task. Furthermore, they should be given any relevant contextual information, such as source files and graphics.

See Chapter 23: Translation Practice: Validation for further information on validation.

 What Does That Mean?

Terminology Database (Termbase)

A collection of standardized terms, often specific to a particular product or industry, used by translators. When these standard terms are encountered in text, their pre-translated versions are automatically inserted by the computer-assisted translation (CAT) tool, ensuring consistent and high-quality translations. Termbases are essential for maintaining term consistency across various translations and are regularly updated and validated for accuracy.

22.6.4 Layout Work

Desktop publishing (DTP) can be a significant cost, so ensure you consider this when deciding on the scope of your translation task. Request that the translation service provider to include this in the quote to avoid any unpleasant surprises later on (see Chapter 23: Translation Practice: Layout Work).

22.7 [Theory] Roles and Responsibilities in the Translation Process

The following table gives an overview of the roles and responsibilities of the main stakeholders in the translation process. These can vary slightly, so the responsibilities defined serve as examples and can be amended to suit your organization.

Table 32: Roles and Responsibilities in the Translation Process

Role	Responsibilities
Technical Writer	- Be mindful of the guidelines on writing for translation when creating content and remember to use internationalization principles. - Ensure all source files (text and graphics) are available for translation. - Act upon any feedback about errors in the source files spotted by translators or validators.
Translation Manager / Translation Responsible	- Decide on translation requirements and scope, and communicate these to the translation partner. - Coordinate product training for translators as required. - Approve or reject quotes from the translation partner. - Process and release the translated documents. - Act on quality issues in cooperation with the translation partner. - Pass on feedback relating to errors in the source text to the technical writer.

Role	Responsibilities
Translation Service Provider / Translation Partner	- Prepare quotes for the translation orders. - Manage the translation and validation process. - Give validators advance warning when a validation task is coming. - Follow up with validators on deadlines. - Ensure the use of company-specific terminology. - Coordinate training of translators as required. - Apply the four-eyes principle to every translation. - Host the company-specific translation memory. - Implement all relevant comments and changes from the validators and ensure they are included in the translation memory. - Deliver the translations by the agreed deadline. - Offer validators training in the use of the validation tool.
Validators	- Check translations. - Check consistency with approved terminology for their language. - Complete the task within the given timeframe. - Use the tool provided by the translation partner to complete the validation task.

22.7 [Theory] Roles and Responsibilities in the Translation Process

 Note

Important Consideration for Validators

Validators should not rewrite content that deviates from the source file or check technical data. Instead, the validators' questions or comments concerning translated content should be communicated to the technical writer via the translation manager.

Chapter 23 Translation Practice

Lead Writer: Alison Pickering | **Peer Reviewer/s**: Kieran Morgan | **Expert Reviewer/s**: Stephanie Riches Harries

Building upon the theoretical foundations in Chapter 22: Translation Theory, this chapter provides practical strategies for managing translation projects in technical documentation. It provides strategies for preparing documents and images for translation, selecting suitable translation service providers, and fostering collaboration to ensure translations maintain accuracy and cultural relevance. Throughout this chapter, we'll guide you through the details of the translation process, from document setup to final delivery, by applying the knowledge gained from our theoretical exploration.

 Who Should Read This

- Technical writers, professionals, and project managers who are responsible for overseeing the translation and localization of technical documentation.

CONTENTS

23.1 Introduction 401
23.2 [Process] Translation and Localization Process 401
23.3 [Practice] Translate and Localize Content 402
23.4 [Practice] Select Translation Partner 408
23.5 [Practice] Create Terminology Database 413

Chapter 23 Translation Practice

PROCESS

Inputs	Translate > Translation Practice		Outputs
• Source Text • Source Images	**Theory**	• Translation and Localization Process • See Chapter 22: Translation Theory.	• Translated and Localized Content
	Practice	• Translate and Localize Content • Select Translation Partner • Create Terminology Database	
	Templates	• Translation Best Practices Checklist for Text and Graphics	

23.1 Introduction

The art of translation extends beyond mere word-for-word substitution. In this chapter, we explore the practical aspects of translation and localization in technical documentation. While the previous chapter, Chapter 22: Translation Theory, laid the foundation by exploring the theoretical underpinnings of translation, here we focus on strategies for effectively managing translation projects.

From understanding how to set up documents and images for translation to selecting the right translation partner, this chapter aims to provide you with practical insights. We will guide you through the translation process from initial document preparation to the final stages of delivery. Emphasis is placed on the importance of working collaboratively with translation service providers and ensuring that your technical content is not only accurately translated but also culturally relevant for your target audience.

Whether you are new to managing translation projects or looking to refine your existing practices, this chapter offers a comprehensive guide to help you navigate the complex landscape of translation and localization with ease and confidence.

23.2 [Process] Translation and Localization Process

This section provides a visual overview of the translation and localization process. It is the foundation on which the steps below are based.

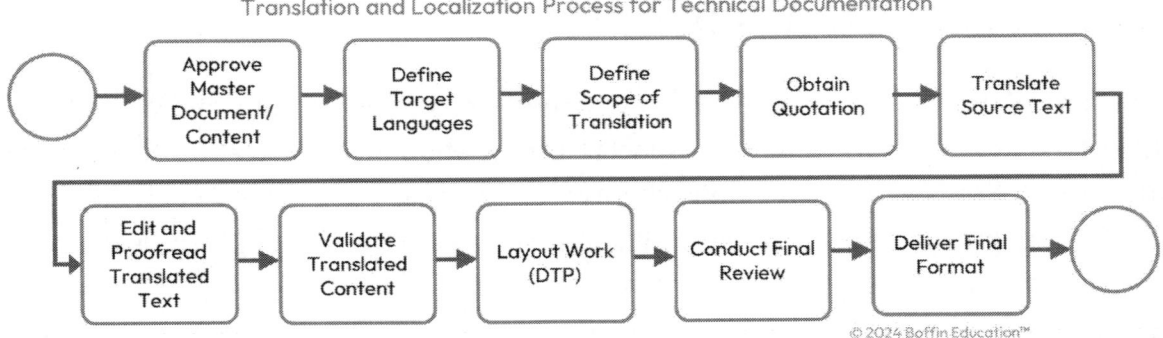

Figure 56: Translation and Localization Process for Technical Documentation

23.3 [Practice] Translate and Localize Content

23.3.1 Approve Master Document/Content

Once the master document/content has been approved—after the final checks and proofreading (see Part 8: Review)—it is ready for translation. The information in this section is based on the assumption that the translation order is sent once the whole technical document is released; however, some organizations choose to release and translate on a chapter-by-chapter or section-by-section basis. This approach can be risky because a change in another chapter during the writing stage may affect a chapter that has already been translated, necessitating retranslation. Consider this when deciding how to order translations.

23.3.2 Define Target Languages

When defining the target languages for the translation of product documentation, it's important to consider target markets, legal requirements, and budget. For example, if the only target market for a product is in Europe, there's no need to translate the product documentation into Chinese.

23.3.3 Define Scope of Translation

At this stage, you'll need to decide on the scope of service required from the translation service provider. Will machine translation with post-editing suffice, or is human translation required? Do you want the translation to be validated? Should the translation service provider carry out any desktop publishing (DTP) work?

All these factors can impact cost and turnaround time and should be given careful consideration. The scope of the translation should be defined and clearly communicated to the translation service provider at the quotation request stage.

23.3.4 Obtain Quotation

Compile all the information relating to the translation task and approach the translation service provider for a quotation. Many translation agencies have an online customer portal for ordering. If not, they will inform you how they wish to receive the files, such as via email or, if they are large files, via WeTransfer, FTP/SFTP, or through a link to the file in Google Drive, Dropbox, or another cloud storage system.

Some companies with advanced business processes will give translation partners access to their content management system (CMS) to enable them to export (pull) the translation source material and import (push) the completed translations.

23.3.5 Quotation Request

When requesting a quote, provide as much information as possible. The translation service provider may have set criteria for what information they require; otherwise, you should include the following information:

- Scope of the translation (level of service and languages required)
- Deadline
- Whether validation is required
- Whether layout work (desktop publishing) is required

- Target audience definition (to enable the translator to adopt the right tone of voice)
- Source text (ask which format they require)
- List of pre-approved terminology (if available)
- Copy of the company style guide (if available)
- Final layout draft (if available)
- Contextual information (e.g., images and training materials)
- List of any product names or technical terms that should not be translated

23.3.6 Quote Format and Content

Once the translation service provider receives your files, they use specialized translation management tools to analyze the files and identify already translated content (100 percent matches or fuzzy matches), as well as any repetitions in the source text. If matches and repetitions are found, you are usually given a discount for them, so you will only be charged the full price for the translation of completely new content.

Quotes are generally given per project on a per-word basis unless requested otherwise, and per-word rates can vary between languages. Any discount given for 100 percent matches, fuzzy matches, or repetitions found in the source text is usually already applied. Desktop publishing costs are usually calculated per hour or per page.

Insight

Types of Text Matches in Translation

100 Percent Matches

100% matches, as the name suggests, occur when the source text is identical to one that has been translated in the past, and editing is not required.

Fuzzy Matches

A fuzzy match is a segment in a source text similar to a segment in a translation memory. It's only a partial match, so it will require editing. The amount of editing needed is reflected in the percentage. For example, a 60% match will probably need complete rewriting, whereas a 99% match may only need a minor change, such as adding a comma.

Repetitions

Repetitions are instances where the same sentence is found more than once in the source text. It will only need translating once, and the translation can then be reused for the subsequent occurrences in the source text.

Utilizing Translation Memory and Terminology Databases

100% matches and fuzzy matches can only be found if the translation service provider has created and been using a translation memory for your organization. You should also have a glossary of key terms, also known as a terminology database (or termbase for short). This ensures that whenever you send new files for translation, the content is cross-checked against your previous translations and key terms, and you are only charged for new content.

23.3.7 Quotation Review and Approval

If the translation service provider has its own platform for requesting translation quotes, they may place quotes there for you to download then notify you via email when they are available. Otherwise, they are usually sent by email.

Deadlines stated as specific dates are usually based on timely approval of the quote. Therefore, if it has taken you several days to approve the quote, it is important to re-confirm the deadline. Some agencies handle deadlines differently, stating them as a specific number of working days after quote approval. The sooner you approve the quote, the sooner you will receive your translation.

23.3.8 Translate Source Text

The translation is typically done with a computer-assisted translation (CAT) tool, which divides the source text into segments, generally full sentences or paragraphs. This segmentation makes it easier for the translator to work on and translate each segment individually. Translation tools help translators speed up their work and enhance accuracy.

Translation may be done by a human translator or with machine translation, as specified in the quotation and confirmed by you during the quotation approval stage.

23.3.9 Edit and Proofread Translated Text

Normally, the editor works in the same translation tool as the translators; however, their task is not translating, but revising each segment.

As described in Chapter 22: Translation Theory: Machine Translation (MT) and Post-Editing, editing and proofreading are conducted on the translated text to ensure correct style and content, and to verify the translation's accuracy. This stage ensures that the translator has accurately rendered the source text, has not omitted any information, and has fully complied with any guidelines provided.

The "four-eyes principle" mentioned in Chapter 22: Translation Theory: Translation, Editing, and Proofreading means that a second language reviewer examines the text. This practice doesn't imply the translator's work is subpar; it is an industry-standard practice based on the idea that two people (thus, four eyes) see more than one person. This is why translations should undergo a quality-assurance review performed by someone other than the translator.

23.3.10 Validate Translated Content

If you have requested validation for the translation, the translation service provider will send it to the designated validator. There is usually a set deadline for this task. The translation service provider will need clear instructions on how to handle situations such as the validator not being able to complete the task within the given timeframe: should they extend the deadline, accept a partially completed validation, or skip the validation? Another likely situation is that the validator asks for clarification on some of the content. In this case, they will need to know who in your organization they can contact for support on that.

23.3.11 Layout Work (DTP—Desktop Publishing)

If desktop publishing is part of the brief you gave to the translation service provider, this stage is where the layout work is undertaken. That means the original text is removed from your source file and replaced with the translation. It applies to all types of content, such as text, tables, warnings, and graphics.

23.3.12 Conduct Final Review

The final review mainly focuses on checking that the translated text sounds natural and reads smoothly in the target language, and on detecting any errors or inconsistencies with regard to punctuation and capitalization. It also ensures that any validation comments have been implemented and that the translation memory is updated accordingly.

If desktop publishing work is included in the translation task, the document formatting is also checked for any issues relating to fonts, images, layout, style, and so on. It also ensures that the finished job conforms to the brief you provided.

The final review stage should be handled by a professional assigned only to that unique task, who has not yet had any contact with the work during previous stages. This way, the reviewer can keep an open mind and has a fresh set of eyes.

23.3.13 Deliver Final Format

The final deliverable format will be as you specified in the brief: Microsoft Word, PDF, XML, HTML, or other format.

Delivery back to you may be via email, or if the agency has a specific portal or platform, they may place the completed translations there for you to download and inform you via email that they are now available.

23.4 [Practice] Select Translation Partner

Of course, you will want to find professional translation services of the highest standards. However, not all translators have the expertise and experience required to work with technical documents, so collaborating with the right language experts is essential. It's not a simple task to identify and select a professional translation service provider, and a lot rides on your decision.

Why are we using the term "translation partner" rather than "translation service provider"? Well, that's simple—successful translations in the world of technical communications are based on a partnership that improves over time. The more work a translation partner does for you, the better they understand your products and your requirements regarding translation style, tone of voice, terminology, and so on. Most translation agencies use a translation memory, so that will continuously be expanding, and any text included in a new order that they have already translated for a previous order will result in a discounted rate.

Therefore, it is not about finding the cheapest quote each time you translate a document but rather about consistency and finding a translation partner that fulfills your needs and is willing to support you on your journey to achieve great quality translations.

23.4.1 Identifying a Professional Translation Service Provider

The following sections list the essential criteria to use when searching for a professional translation service provider that best suits your needs. Look out for these key points to help you decide if they are a professional company that focuses on delivering quality translations.

23.4.2 Certification

Any reliable company offering translation services must be able to provide certifications; otherwise, it would be best to try elsewhere.

ISO 17100:2015 is an international standard published in 2015, which was based on the BS EN 15038 European Quality Standard. It was reviewed five years later, in 2020, by the ISO, which confirmed the version remains current. The ISO 17100 standard details the requirements that a translation services provider (TSP) shall meet in order to provide a translation service that meets the client's and other applicable specifications. It includes provisions concerning the management of core processes, minimum qualification requirements, the availability and management of resources, and other actions necessary for the delivery of a quality translation service. It does not apply to the use of raw output from machine translation plus post-editing.

The four key clauses of the ISO 17100 standard are:

- **Human Resources**: ensuring the people selected to perform translation tasks have the required competencies and qualifications.
- **Preproduction Process and Activities**: ensuring processes are in place for handling and analyzing enquiries, determining project feasibility, preparing quotations, and entering into agreements.
- **Production Processes**: ensuring compliance with the agreement from the moment it is confirmed to the agreed end of the project.
- **Postproduction Process**: ensuring processes are in place for handling client feedback, assessing client satisfaction, and making appropriate corrections, or taking corrective action.

ISO 9001:2015 covers the requirements for a quality management system and continuous improvement, so any translation service provider with this certification demonstrates their dedication to delivering a high-quality service.

23.4.3 Experience

Experienced translation agencies will have developed their skills and expertise over a number of years, contributing to their reliability, competence, and professionalism.

An experienced translation service provider will only work with native speakers who are subject matter experts. This ensures that your translated text will be high quality, read fluently, and reach your target audience in words they understand. Their expertise will also ensure that your documents comply with the rules and regulations of your target market.

If the translator is unfamiliar with the specific industry, its requirements, and its specific terminology, the translated content will lack precision and accuracy, and in the worst-case scenario, might be mistranslated. Such mistakes could have dire consequences for your business in terms of costs, litigation, and reputation.

23.4.4 Communication

The way a company communicates is one of the most important factors to consider when choosing your translation partner. To avoid misunderstandings, they should clearly communicate how they select their linguists, what is included in the price, and their actual translation process.

They should also be upfront and honest about their security measures, including data protection and any data privacy guarantees. If you need a nondisclosure agreement, ask whether they will provide one or be willing to sign yours.

When dealing with professional translation agencies, you should be able to speak freely with the project manager about any queries or concerns. The bottom line is that they should be open, honest, and inspire confidence in their skills and services.

23.4.5 Tools and Processes

The tools a translation service provider uses and their approach to completing your translation are good indicators of the quality level they can offer. Therefore, ask the translation service provider to detail its translation process and list the tools it uses. A professional translation service provider will use computer-assisted translation (CAT) tools, as these support the creation and use of translation memories and terminology databases (see Chapter 54: Translation Practice: Creating a Terminology Database for Industry-Specific Terms).

23.4.6 References

Any reputable translation service provider will be willing to provide references from current customers as well as examples of their work.

23.4.7 Test Translations

A translation service provider eager to earn your repeat business will be willing to complete test translations. Ensure that any test translation texts you provide are similar to the content they will be required to translate in the future, and have a native speaker with the necessary technical knowledge review them.

23.4.8 Agreeing Terms

As part of building ongoing cooperation with a translation service provider, consider agreeing to terms relating to:

- Time taken to provide quotes (e.g., three working days).
- Time taken to translate (e.g., 20 working days).
- Time taken for validation (e.g., 15 working days).

23.4.9 Handling Questions and Queries from the Translation Service Provider

Good communication between the translation service provider and the translation responsible at your company is vital. Questions and queries from the translation service provider should be answered as soon as possible, as a delayed response could result in the agreed deadline being pushed back.

Examples of questions/queries include:

- Translators or validators may require clarification on ambiguities in the source texts.
- Some source files or graphics may be missing.
- Validators may request a longer deadline to complete the validation task, and the translation service provider needs to know whether you agree to this.

23.4.10 Learnings

Any feedback from the translation service provider relating to the translation task should be passed on to the relevant stakeholder in your company. For example, any errors found in the source text should be passed on to the technical writer.

23.5 [Practice] Create Terminology Database

Terminology databases, also known as glossaries or termbases, consist of a set of standard terms for translators to use. This is particularly useful if you use a lot of product- or industry-specific terms. Once a terminology database is created, each time the translation tool spots one of the standard terms, the translated version automatically appears in the CAT tool. This contributes to achieving consistent translations.

How to Create a Terminology Database:

1. Create the source terms (such as key terminology, brand names, industry-specific terms).
2. Get the translation service provider to translate the terms into the languages you require (you may need to answer queries relating to exact meanings and usages during this stage).
3. The designated validators validate the translated terms for their language.
4. The translation service provider makes any adjustments resulting from validation, then saves the terms in the terminology database for future reference.
5. Repeat the process at regular intervals (for instance, on an annual basis or when a certain number of new terms have been added).

 Note

Agency-Assisted Terminology Database Creation

Some translation agencies offer services to create a terminology database for you by extracting key terms from the source files and then discussing their relevance with you. This would replace step 1 in the creation process.

Part 10 Publish

Navigating the final step: Publishing with precision and control.

Chapter 24 Publish

Lead Writer: Kieran Morgan | **Peer Reviewer/s**: John New

This chapter focuses on the final stages of document creation, emphasizing document control and publication. It guides technical writers through principles and practical steps such as establishing document control, obtaining approvals, conducting final checks, and communicating with stakeholders. Additionally, it distinguishes between documents and records, explores the document lifecycle, and aligns its principles with ISO 9001 standards.

 Who Should Read This

- Aspiring Technical Writers
- Beginner Technical Writers
- Cross-Domain Professionals

CONTENTS

24.1 Introduction 419
24.2 [Theory] What Is a Document? 420
24.3 [Theory] Document Lifecycle 421
24.4 [Theory] Document Control 423
24.5 [Practice] Publish Final Version 430

Chapter 24 Publish

PROCESS

Inputs	Publish > Publish		Outputs
• Reviewed Final Draft	**Theory**	• What Is a Document? • Document Lifecycle • Document Control	• Request for Approval • Controlled and Published Document • Message to Stakeholders
	Practice	• Establish Document Control • Obtain Approval • Conduct Final Checks • Publish Final Version • Communicate with Stakeholders	
	Examples	• Sample Document Control Table • Sample Message to Stakeholders Announcing Publication • Sample Request for Approval	
	Templates	• N/A	

24.1 Introduction

Imagine reaching the summit of a mountain after a long and arduous climb. The moment you stand at the top, you savor the panoramic view and bask in the triumph of your journey. You allow yourself to relax for just a moment, knowing you'll need your energy for the next stage.

In technical writing, publication is a similar moment. When you publish your document, it represents the summit of your successful journey through the Technical Writing Process. It's a moment to be savored! It's where your documents transition from being cost centers—consumers of resources and effort—to value adders, creating value for users and making all your efforts worthwhile.

This chapter will guide you through this transformative process, emphasizing the importance of the document lifecycle and effective publication practices. You'll learn how to establish robust document control, ensuring that your documents are well-managed and compliant with industry standards. We'll explore how to effectively secure approvals, ensuring that your document meets the necessary criteria before publication.

Additionally, this chapter addresses the critical steps of finalizing your document. This includes conducting thorough final checks and refining your document to perfection. We also emphasize the importance of clear and strategic communication with stakeholders, ensuring that the transition of your document from a draft to a published work is smooth and well received.

This chapter aims to equip you with the knowledge and tools to make publication a straightforward process, reflecting the success of your journey through the Technical Writing Process.

24.2 [Theory] What Is a Document?

Before we dive into the nuts and bolts of document control, let's start by explaining some basic concepts. We'll discuss what documents are and how they're closely related to—but distinct from—records. This is a fundamental distinction in document control:

- **Documents** are forward-looking pieces of information used to guide decision-making or describe what should be done. In technical writing, they typically provide task-oriented information that enables an internal or external audience to accomplish a goal.

- **Records** are historical information, evidence of past activities and decisions. They serve as the organizational memory of what was done. They're a static snapshot of something that happened in the past, and unlike documents, they usually don't change.

 What Does That Mean?

Document

A discrete unit of information used to guide work, decisions, or judgment, serving as a guide to what should be done. Documents are forward-looking, as opposed to records, which are historical. Examples include technical documentation, plans, policies, and engineering drawings.

Record

A discrete unit of information or collection of data that forms evidence of past activities or decisions, serving as a static memory of what was done. Records are historical, in contrast to documents, which are forward-looking. Examples include invoices, test results, and completed maintenance logs.

24.3 [Theory] Document Lifecycle

The document lifecycle constitutes the macro stages that all documents go through in their lifecycle.[51] Think of it as encompassing the entire lifespan of a document: from metaphorical birth to death.

In the diagram below, we've created our own version of the document lifecycle, drawing on Kassa's framework in *Document Control: Lifecycle and the Governance Challenge*.[52] We've aligned the terminology more closely with ISO 9001, a key standard relating to document control, which we'll discuss shortly.

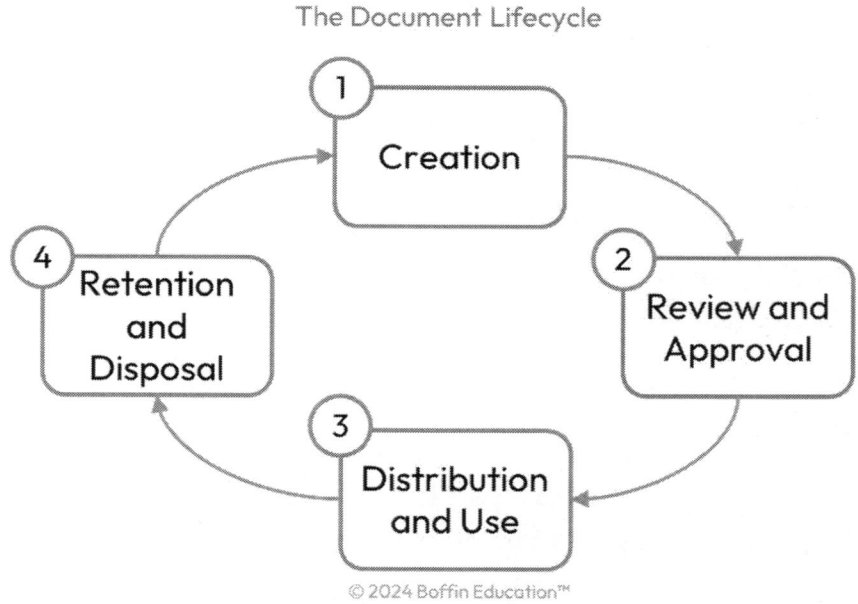

Figure 57: The Document Lifecycle

What Does Each Stage Involve?

1. **Creation**: In this stage, documents are born. They may exist initially as a sketch of a document outline on a notepad or as rough notes in an electronic file before progressing to a fully fleshed-out draft.

2. **Review and Approval**: This stage concerns the review of documents to ensure they're accurate and fit for purpose, and to show that they've been endorsed for use by the appropriate authorizers within an organization.

3. **Distribution and Use**: This is where documents are used by internal or external users to accomplish a goal, such as carrying out a procedure or using a software application. Before this can occur, the documents must be controlled.

4. **Retention and Disposal**: In the final stage of their lifecycle, documents are reviewed for relevance according to an organization's policies and disposed of if necessary, including the removal of any outdated copies.

 What Does That Mean?

Document Lifecycle

The macro stages of a document as it progresses through its lifespan, from metaphorical birth to death, including creation, review and approval, distribution and use, retention, and disposal.

 Insight

ISO 9001—The Elephant in the Document Control Room

When discussing document control, it's hard to avoid the standard for quality management systems, ISO 9001. This international standard is widely used in many industries, particularly by companies bidding for government contracts. One of the processes defined in ISO 9001 is document control. Here, many of our document control principles are embodied as requirements, although in much less detail than our definitions. Our document control principles align closely with ISO 9001:2015, section 7.5—Documented Information.

24.4 [Theory] Document Control

Document control is the process of managing information through the phases of the document lifecycle. Although it doesn't take much time to apply, it's an incredibly important moment. It's the mechanism that enables documents to be managed over their lifespan. Without effective document control, an organization's information can quickly descend into chaos. This can cause problems such as defective products, inconsistent output, duplication of effort, customer complaints, costly calls to helpdesks that could have been diverted, and corruption of the organization's institutional knowledge.

The key principles of document control are illustrated in the diagram below. These principles draw from the ISO 9001:2015 standard, which are explained in *ISO 9001:2015 in Plain English*,[53] and general principles defined by *Document Control: Lifecycle and the Governance Challenge*.[54]

We've drawn on these sources to identify the principles that are your responsibility as the technical writer, which we explain in detail in this section. The others are (mostly) the responsibility of the document controller or knowledge manager—the roles accountable for maintaining the integrity of the overall document control system. These principles aren't discussed in depth, as they're beyond the scope of this book.

Chapter 24 Publish

The Principles of Document Control

Creation	Review and Approval	Distribution and Use	Retention and Disposal
✓ Identification ✓ Format and Media	✓ Review ✓ Approval	✓ Availability and Accessibility ✓ Classification and Categorization ✓ Access Control ✓ Information Protection ✓ External Document Control*	✓ Version Control ✓ Change Control ✓ Protection from Alteration* ✓ Storage and Preservation* ✓ Retention Guidelines* ✓ Disposal and Disposition*

*Primary responsibility of the Document Controller or Knowledge Manager, not the Technical Writer.

Figure 58: The Principles of Document Control

 What Does That Mean?

Document Control

The process of managing documents through the phases of their lifecycle.

Controlled Document

A document that requires control throughout its lifecycle. It features a unique identifier and approval details. Using an incorrect version of this document could lead to quality, safety, or compliance issues.

24.4.1 Principle 1: Identification

What It Is: The unique identification of a document that distinguishes it from other documents.

This principle facilitates a single source of truth for documents, allowing for traceability back to the source document and the coordinated management of multiple versions. Identification is usually accomplished by an alphanumeric document identifier (ID) generated by a content management system. However, this doesn't necessarily have to be the case. It can be any unique aspect that allows you to differentiate between documents.[55] The identifier should be clearly marked on the document to ensure traceability.

24.4.2 Principle 2: Format and Media

What It Is: Selection of the appropriate media (e.g., print, digital, or both) and format (e.g., web page, Adobe PDF file, Microsoft Word file, etc.) for a document.

This principle involves selecting the appropriate format and media for your audience. It includes integrating images to aid in usability and comprehension, and translating documents into a language suitable for your audience. This decision should have been made well before publication, as part of your planning and audience analysis. Your organization may already have well-defined guidelines and templates for the appropriate documentation format that you must follow. We discuss audience analysis in depth in Chapter 11: Analyze Audience and translation in Part 9: Translate.

24.4.3 Principle 3: Review

What It Is: Review of a document by experts to ensure quality, accuracy, and suitability for use by the intended audience.

Review is an essential aspect of the document lifecycle and the Technical Writing Process. We discuss review extensively in Part 8: Review, so we won't go into detail here, beyond this brief explanation.

> **What Does That Mean?**
>
> **Review**
>
> The process of evaluating a document against quality standards such as technical accuracy, consistency with style manuals, templates, branding, and so on.
>
> **Reviewer**
>
> An expert responsible for evaluating a document against quality standards such as technical accuracy, consistency with style manuals, templates, branding, and so on.

24.4.4 Principle 4: Approval

What It Is: Endorsement of a document for publication by approvers or document owners.

Approval is a cornerstone principle of document control. It's the final step in the review process that indicates a document is fit for publication. For information on how to obtain approval, see Chapter 24: Publish: Obtain Approval.

What Does That Mean?

Approval

The formal acknowledgment or sign-off by an approver or document owner that a document is fit for publication.

Approver

An authorizing person within an organization who confirms that a document is fit for publication.

Document Owner

A manager or delegate responsible for ensuring a document is current and accurate.

24.4.5 Principle 5: Availability and Accessibility

What It Is: Ensuring documents are available and accessible to those who need to use them.

Documents need to be distributed to users in a location where they can easily locate and navigate to find the information they're looking for. In technical writing, this is usually accomplished by uploading a document to the organization's knowledge base or content management system and applying the correct metadata so that users (internal or external) can easily find them. Usually it's the responsibility of the document controller or knowledge manager to oversee the content management system so that it's accessible.

24.4.6 Principle 6: Classification and Categorization

What It Is: The categorization of documents according to your organization's information taxonomy.

Documents can be categorized in myriad ways—by department, by topic, by product line, and so on. This allows documents to be easily discovered through search. This categorization is typically not visible on the document itself; it's usually stored as metadata in a content management system. Sometimes, a document's classification is visible on the document if it's helpful for users to know.

> **What Does That Mean?**
>
> **Metadata**
>
> Data about data. For example, in digital photography, metadata captured alongside a digital image typically includes the date and time the photo was taken, the brand and model of the camera used, and the geographical location it was taken in.

24.4.7 Principle 7: Access Control

What It Is: Ensuring that only those who should have access to your document do.

This principle maintains the confidentiality of sensitive information and avoids improper use. Typically this is accomplished via the use of restricted areas within organizational knowledge bases accessible only to authorized users. It can also be accomplished manually using security classifications and document distribution matrices, ensuring that sensitive documents reach only their intended audience.

24.4.8 Principle 8: Version Control

What It Is: Identification of the correct version (or "revision status") of your document.

This is an essential element of document control as it enables users to know if they are using the current version of a document. Often, but not always, version numbers are allocated automatically by content management systems. This number should be clearly marked on every document, alongside the alphanumeric identifier, to provide visibility and enable traceability. More details on versioning are discussed in Chapter 17: Write Draft: [Theory] Version Control.

24.4.9 Principle 9: Change Control

What It Is: Identification of changes from one version of your document to the next.

This principle is especially important for internal documentation where users need to be able to view changes to documents so they can understand how to revise their work practices. For external documents, it's more a matter of good housekeeping. It keeps writers and approvers abreast of the differences from one version to the next, who authored them, and the date they were approved and published. Changes won't usually be visible on documents with an external audience, unless it has implications for how they conduct their work. Typically, change control and versioning are facilitated automatically via a content management system, which will automatically compare one version to another.

 What Does That Mean?

Internal Documentation

Technical content for an organization's internal audience, focusing on technical details, processes, and operational efficiency.

External Documentation

User-oriented content for customers and partners outside the organization, emphasizing ease of use, product understanding, and brand representation.

24.4.10 Principle 10: Information Protection

What It Is: Locking of finalized documents to prevent unauthorized changes.

Information protection often involves the publication of a document in a noneditable format, such as a web page only editable by authorized internal users, or the distribution of a secured digital version, such as an Adobe PDF file. Consult your organization's document control policy to see what its expectations are.

24.5 [Practice] Publish Final Version

The publication of a final version occurs after you've completed the steps in the Technical Writing Process: Plan, Design, Write, Edit, and Review. By this stage, your documents should be highly polished, reviewed, technically accurate, and well formatted using your organization's template or branding

24.5.1 Step 1: Establish Document Control

The first step in publication is to establish document control. For most documentation, that's straightforward: ensure your document includes a document control table that's ready to be populated with relevant data. If your organization employs a sophisticated content management system, document control might be integrated into the workflow to automatically establish version control and record details and dates of approvals. If you're using a document control table and your organization lacks a standard template, we've created a Sample Document Control Table you can modify for your documents.

24.5.2 Step 2: Obtain Approval

Obtain a record of approval from approvers or document owners. To do so, you can send a Request for Approval via email or use the workflow functionality built into your organization's content management system or help authoring tool. To help you craft a Request for Approval, we've supplied a Sample Request for Approval you can customize.

 Insight

Always Keep a Record of Review and Approval

When you're going through review and approval, it's essential to keep records. Don't ever rely on a verbal approval! Not only is this poor document control—and probably against your organization's policies—it's risky for your career. If something goes wrong and an error is found in your document, there'll most likely be an investigation to trace the error back to the source. If it looks like you've allowed documents to be published without sufficient review and approval, that'll reflect poorly on your technical writing skills and perhaps result in disciplinary action. If someone gives you verbal approval, follow it up with a request for written confirmation.

24.5.3 Step 3: Conduct Final Checks

Once your document has been approved, it really shouldn't change much from this point on. In fact, some organizations have strict content management systems that won't allow changes after approval without sending the document back to an unapproved state. This step is about giving your document a final once-over for any layout and formatting issues that may have cropped up between versions.

Chapter 24 Publish

We suggest you use our Editing Checklist Template to perform final checks. You don't need to do all the checks again, but we recommend you complete the Layout and Format Checking steps, and also conduct a quick proofread to ensure your document is as polished as possible. If your document will be distributed as a printed guide, refer to the Integrity Checking section in the template.

 Insight

Use a Printout for a Fresh Perspective

It often helps to print a document, even if it's only going to be published electronically (remember, your audience may want to print it out too). If you've gotten used to reading it on a screen, looking at a printed copy can give you a fresh perspective, enabling you to spot errors you may have overlooked.

24.5.4 Step 4: Publish Final Draft

Ensure your document has the correct classification and categorization metadata applied. If required by your organization's policies, protect it from alteration. When you're all set, upload and publish your document on your organization's content management system or knowledge base, or follow the production steps defined by your organization for printed guides.

 Tip

Protecting Files

Protection might be inherent in your organization's publishing platform—for example, through an online platform that only authorized users can access. However, documents are often distributed as files. In these cases, protection might involve saving the document as a password-encrypted, noneditable version.

24.5.5 Step 5: Communicate with Stakeholders

Finally, once your document has been published, it's good practice to follow up with a brief announcement, particularly if it's an important document. This step might not always be required—for example, if notifications are built into your organization's content management system or if it's a routine document such as a work instruction that doesn't need much fanfare.

Often a simple email to the relevant stakeholders will suffice. Consult your documentation plan for your list of stakeholders, and ensure you send it to those affected by the document's publication, such as document owners and managers responsible for distributing new and updated documents to their teams. We've crafted a Sample Message to Stakeholders Announcing Publication you can customize.

Part 11 Manage

Methods to effectively manage technical writing projects.

Chapter 25 Manage Progress

Lead Writer: Kieran Morgan | **Peer Reviewer/s**: Amanda Butler

This chapter provides a comprehensive guide for tracking and managing progress in technical writing projects. It explores various tools and methodologies, including checklists, status trackers, visual management boards, and project schedules, offering practical advice on their application. The chapter equips writers and managers with strategies to efficiently manage milestones, adapt to changes, and ensure timely completion of technical documentation tasks.

 Who Should Read This

- Beginner Technical Writers
- Career Advancers
- Managers of Technical Writers
- Project Managers
- Cross-Domain Professionals
- Consultants

CONTENTS

25.1 Introduction 439
25.2 [Theory] Checklists, Status Trackers, Visual Management Boards, and Schedules 441
25.3 [Practice] Manage Progress 452

Chapter 25 Manage Progress

PROCESS

Inputs	Manage > Manage Progress		Outputs
<ul><li>Documentation Plan</li><li>Project Schedule</li><li>Technical Writing Process</li></ul>	**Theory**	<ul><li>Checklists, Status Trackers, Visual Management Boards, and Schedules</li></ul>	<ul><li>Checklist</li><li>Status Tracker</li><li>Visual Management Board</li><li>Updated Project Schedule</li></ul>
	Practice	<ul><li>Define Writing Process</li><li>Break Down Scope</li><li>Choose Progress Tracker</li><li>Set Up Progress Tracker</li><li>Populate With Tasks</li><li>Manage Progress</li><li>Manage Issues</li><li>Report Progress</li><li>Analyze Data</li></ul>	
	Examples	<ul><li>Sample Documentation Project Schedule</li></ul>	
	Templates	<ul><li>Simplified Technical Writing Process Checklist</li><li>Technical Writing Process Checklist</li><li>Status Tracker Template</li></ul>	

25.1 Introduction

Imagine steering a ship through a misty sea without a compass—that's what tackling a technical writing project without tracking progress can feel like. In the ocean of tasks and deadlines, it's easy to lose sight of your destination.

This chapter is your compass. From the chaos of unmanaged tasks to the calm of a well-organized project, we'll guide you through the essentials of managing progress. Whether you're juggling multiple documents or focusing on a single, critical deliverable, the tools and techniques presented here—checklists, status trackers, visual management boards, and project schedules—will illuminate your path. You'll stay on course, meet your deadlines, and keep your sanity intact.

Whether you're a solo writer tracking to someone else's deadline or a senior writer managing a complex project, this chapter will be valuable. It will equip you with the knowledge and skills needed to navigate the waters of technical writing project management with confidence.

Chapter 25 Manage Progress

 **Tip**

Integrate Your Workflow with the Project Team

As technical writers, we're often part of larger projects, making it essential to align with the project team's workflow and tools.

Consider these examples:

- You're part of an Agile software development project where the team uses issue tracking software for development. Align with them by using the same software for tracking your documentation progress. Participate in sprint meetings and daily stand-ups to stay informed and add your input.

- Your project manager organizes regular status meetings to keep the project schedule up-to-date. Make sure you're a consistent attendee. Keep them informed about your progress and any potential delays in meeting documentation milestones.

Being integrated with the project team's tools is vital. This brings more visibility to your efforts and ensures they're in line with the overall project objectives.

25.1.1 When Do I Start Managing Progress?

Managing progress underpins all other activities in the Technical Writing Process—it's an ongoing activity throughout the duration of your project. This is why we depict it along the bottom rung of the process diagram in Chapter 7: Tailor the Process.

It's an essential step in any well-run documentation project, but it's something you can't begin before you've done your homework. Before managing progress, you'll need to set up the framework so you can do so effectively. This revolves around two activities: defining your high-level workflow and breaking down the scope of your project into documents, or "deliverables."

These two activities go hand-in-hand:

1. **Structuring Workflow**: Defining your writing process helps you understand the macro phases of your workflow, creating a structured framework for each topic and task to progress through, such as Design > Write > Edit > Review > Approve > Publish. You'll use these high-level phases as the stages in your progress tracker. For more information on how to do this, see Chapter 7: Tailor the Process.

2. **Define Deliverables**: Breaking down your project's scope into more detailed deliverables is an essential step in managing progress. It allows you to view your project as a series of smaller chunks of work, which can be marked off in your progress tracker one by one, giving you a more accurate view of progress. For more guidance, see Chapter 10: Make a Plan.

 Tip

If you feel ready to tackle more advanced project planning techniques, such as estimating—which gives you a structured approach to break down your project's scope in more detail—check out Chapter 7: Estimate Scope, Time, and Cost in our forthcoming title, *Project Management for Technical Writers*, on our website: https://boffin.education. Use the coupon code in Templates to claim your one-year free subscription to the site.

25.2 [Theory] Checklists, Status Trackers, Visual Management Boards, and Schedules

This section introduces some straightforward tools you can use to manage progress. For each tool, we've provided a template or example: Process Checklist, Status Tracker Template, and Sample Documentation Project Schedule.

Read the sections below to learn more about each tool.

25.2.1 Checklists

Checklists are a simple yet effective way to manage progress for an individual document or topic. They consist of an itemized list of tasks or checks that each document must undergo, typically organized under various headings. Once all checks are complete, the checklist can be moved from an "In Progress" to a "Done" state.

Pros	Cons
• **Simplicity**: Straightforward and easy to use. • **Efficiency**: Help in quickly identifying completed and pending tasks. • **Clarity**: Provide a clear view of what needs to be done, reducing the chance of missing steps.	• **Oversimplification**: May oversimplify complex tasks, leading to insufficient planning. • **Checklist fatigue**: Frequent use can lead to complacency, where items are checked off without proper attention.

How Does It Work?

We've created a Simplified Technical Writing Process Checklist for a single document or topic, following the phases of the Technical Writing Process. If you're looking for something more detailed, check out our comprehensive checklist. It includes detailed tasks and deliverables.

25.2 [Theory] Checklists, Status Trackers, Visual Management Boards, and Schedules

 Insight

From Paper Trails to Digital Footprints

Once upon a time, completing checklists was a physical task, involving moving a checklist from an in tray to an out tray (yes, an actual wooden or plastic tray!). The presence of a checklist in the in tray signaled to the team that new work was incoming. Moreover, the size of the in tray served as a clear indicator of the team's backlog size. These days, many teams—especially those comprising knowledge workers—have shifted to using virtual systems.

25.2.2 Status Trackers

At its core, a status tracker is a straightforward yet effective tool for keeping an eye on the progress of your writing tasks. Usually set up as a spreadsheet, it lets you monitor various stages of your documents or deliverables. It's perfect for those who like their project management tools simple and to the point.

Here's an example of a status tracker in action, using our Status Tracker Template:

Deliverable Name	Topic	Doc ID	Responsibility	Due Date	Milestone	% Complete	Status
Document 1	Topic 1	DOC-1237	Lead Writer 1	Wed 04/9/25	Drafting in Progress	22%	At Risk
Document 1	Topic 2	DOC-1237	Lead Writer 1	Tue 04/22/25	Structure Confirmed	11%	On Track
						0%	
						0%	
						0%	
						0%	
						0%	
						0%	
						0%	
						0%	
Total						17%	

Figure 59: Sample Status Tracker

Pros	Cons
• **Ease of Use**: User friendly and easy to set up. • **No Special Software Required**: Can be created with common spreadsheet applications. • **Team Accessibility**: Shareable with team members for collaborative updating. • **Customizable**: Can be tailored to fit the specific needs of your project.	• **Lacks Automated Features**: No built-in workflow routing or notification systems. • **Manual Updates**: Requires regular manual updating to maintain accuracy. • **Limited Scalability**: May become cumbersome for very large projects with numerous tasks.

How Does It Work?

Status trackers work by assigning statuses or percentages to various stages of your project. In the example below, we've used milestones from the Technical Writing Process to monitor the progress of deliverables such as documents and topics.

% Complete: Documentation

Milestone	% Complete
Design	
Structure Confirmed	11%
Write	
Drafting in Progress	22%
Final Draft Complete	33%
Review	
SME Review Complete	44%
Peer Review Complete	56%
Final Draft Updated	67%
Edit	
Final Draft Edited and Proofed	78%
Publish	
Final Draft Approved	89%
Document Published	100%

Figure 60: Sample Milestones and Percentage Complete in a Documentation Status Tracker

As well as tracking percent completion, the risk to the delivery of each milestone can be visually represented by assigning a "RAG" (red, amber, or green) status. This provides a clear visual cue indicating whether a deliverable is on track, or if it has encountered issues and requires closer management or escalation.

Status

Status	Color
On Track	●
At Risk	○
Critical	●

Figure 61: Sample RAG Status in a Documentation Status Tracker

25.2.3 Visual Management Boards

As the name suggests, visual management boards provide a more visual way to track progress and manage tasks, making them a great choice for managing writing projects. Though they require a little more effort to set up than a status tracker, they offer numerous advantages. These include improved visibility, easier collaboration within a team, and the satisfying sense of accomplishment you get when moving cards to a "Done" state.

These boards originate from methodologies like Lean and Agile. They're driven by the need to reduce waste, improve workflow, and enhance team collaboration. You might have heard of them already—common types are kanban boards and sprint boards. They can be physical or digital, with tools like Trello or Microsoft Planner offering online platforms that are easy to use. These options are affordable and scalable, suitable for small teams and start-ups as well as large organizations.

Pros	Cons
- **Visual Clarity**: Offers a clear overview of the project's current status at a glance. - **Flexibility**: Easily adaptable to changes in project scope or priorities. - **Enhanced Collaboration**: Facilitates team communication and coordination among team members. - **Focus on Efficiency**: Helps in identifying and reducing workflow bottlenecks.	- **Potential Oversimplification**: May not capture complex dependencies between tasks. - **Requires Discipline**: To be effective, it must be kept up-to-date.

25.2 [Theory] Checklists, Status Trackers, Visual Management Boards, and Schedules

How Do They Work?

Visual management boards use columns to represent workflow stages and "cards" to represent tasks. Tasks are moved from one column to another as they progress to completion.

Below is an example of one particular type called a kanban board. This sample uses the phases of the Technical Writing Process as its workflow stages and topics in a technical document as cards.

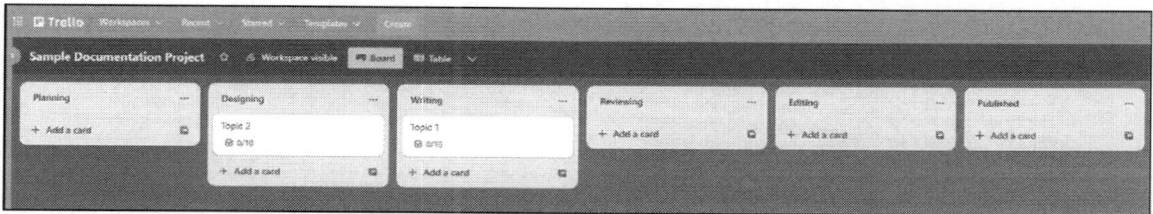

Figure 62: Sample Kanban Board Using the Phases of the Technical Writing Process

Each card on a visual management board can include more detailed steps, facilitating a rigorous quality-assurance process. The sample card below shows a hypothetical sequence of tasks—effectively a topic-by-topic checklist—for developing a single topic, from planning through to publication.

Chapter 25 Manage Progress

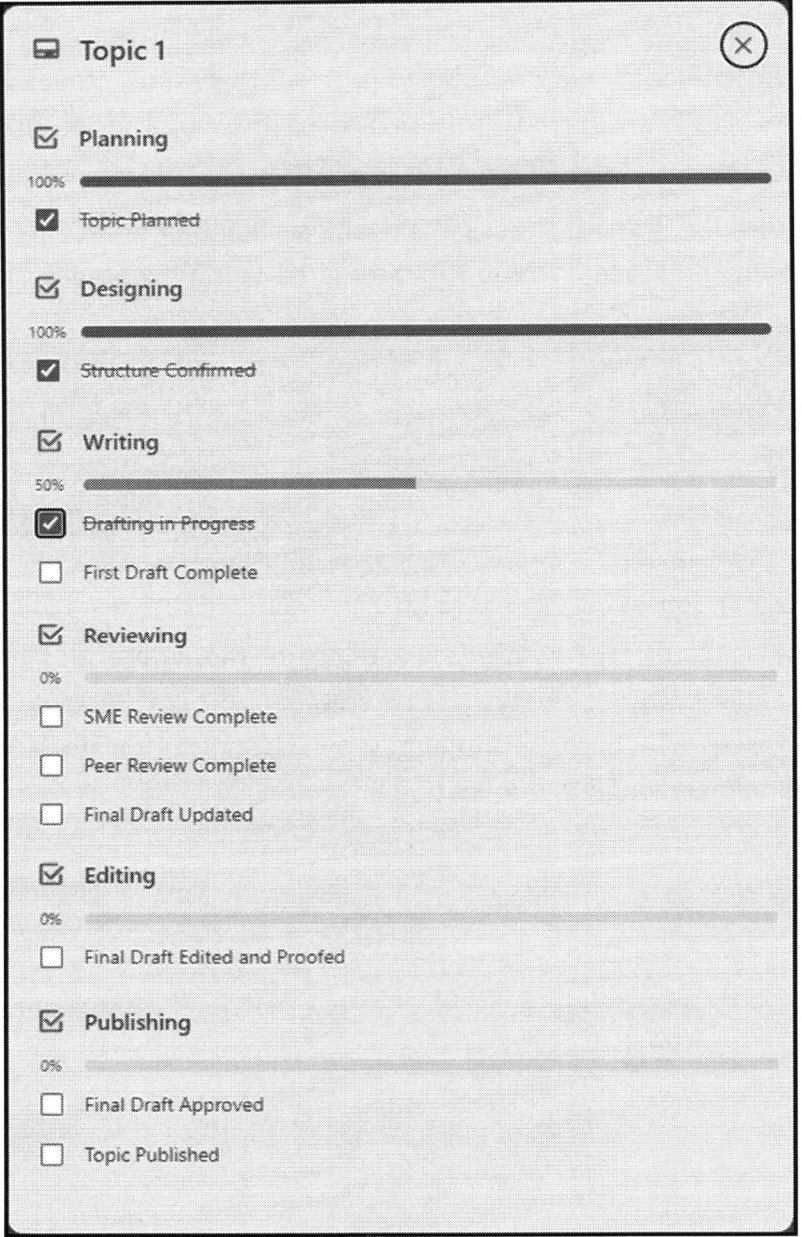

Figure 63: Example Kanban Card with Checklist

25.2.3.1 Kanban Boards vs. Sprint Boards in Technical Writing Projects

Kanban boards and sprint boards are related but uniquely distinct variations on visual management boards. The main difference lies in how tasks are moved and managed, rather than in the use of the visual tool: kanban focuses on a concept called continuous flow,[56] while sprint boards operate within the defined timeframe of sprints.

- **Sprint boards**: These are typically used in Agile project management and center around short, time-boxed periods called sprints, where specific tasks are completed.
- **Kanban boards**: These are more flexible, focusing on continuous delivery without the time constraints of sprints.

Kanban Boards	Sprint Boards
Kanban, from the Japanese word for "signboard," was developed in the 1950s by Taichi Ohno of Toyota to enhance manufacturing processes.[57] It's particularly effective for projects that evolve and require flexibility. Kanban boards visually represent work at various stages using columns and cards. They focus on the prioritization of tasks, allowing for a continuous flow and adjustment based on current priorities. This method is excellent for promoting collaboration, tracking progress, and identifying bottlenecks. Kanban is particularly suitable for projects where tasks evolve over time and where deadlines are flexible.	Central to Agile project management, sprint boards revolve around sprints—short, focused periods of work aimed at completing specific tasks. Originating in software development, they are now widely applied in numerous fields for managing time-sensitive projects. Sprint boards encourage a structured, goal-oriented approach, where tasks are tackled intensively within the defined timeframe of a sprint. This method fosters a sense of urgency and focus, making it ideal for projects with clear goals and tight deadlines. Sprint boards are effective for ensuring rapid progress and adaptability in fast-paced, collaborative environments.

25.2.4 Schedules

If you've built a project schedule and feel confident in your project management abilities, schedules can be a powerful tool for tracking a project's overall progress. They're especially valuable in managing complex projects where the workflow has numerous dependencies and involves multiple team members.

25.2 [Theory] Checklists, Status Trackers, Visual Management Boards, and Schedules

Pros	Cons
• **Detailed Oversight**: Allows for tracking detailed aspects of project progress, ensuring nothing is overlooked. • **Predictive Planning**: Aids in forecasting the impact of task evolution on deadlines for proactive management.	• **Complexity**: Maintaining schedules can be challenging, particularly for large projects with numerous elements. • **Time-consuming**: Regular updates to the schedule can be demanding in terms of time and effort.

How Does It Work?

To help you get started, we've provided a customizable Sample Documentation Project Schedule for you to tailor to your project's needs. Managing progress with a schedule revolves around consistently updating tasks and milestones as they're completed. It's also important to stay adaptable, introducing new tasks and milestones as your project progresses. This helps in accurately forecasting how changes might affect your deadline and lets you proactively address potential delays.

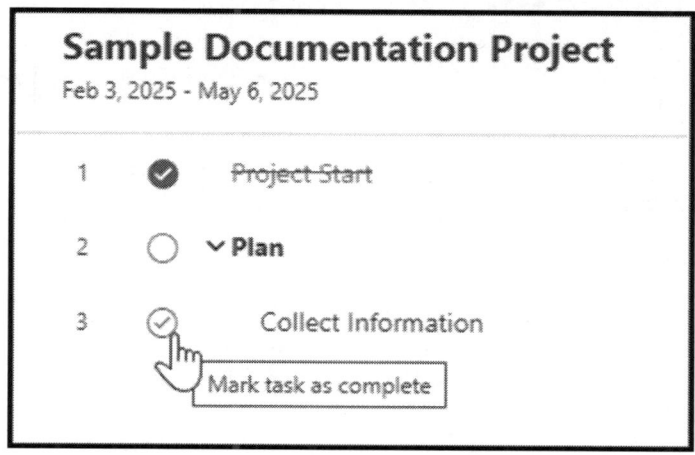

Figure 64: Marking Tasks Complete in a Project Schedule

> **Tip**
>
> If you feel ready to tackle more advanced project planning techniques, such as scheduling, check out Chapter 14: Develop Schedule in our forthcoming title, *Project Management for Technical Writers*, on our website: https://boffin.education. Use the coupon code in Templates to claim your one-year free subscription to the site.

25.2.5 Advanced Options

For those of you ready to dive deeper, there's a world of complex options out there for advanced users. If you're familiar with these systems and have the time to invest, you can create much more intricate workflow management systems than what we've covered. Platforms like Atlassian Jira and Smartsheet provide seasoned users with a range of features, including customizable workflows, automated notifications and reminders, role-based permissions, and more.

25.3 [Practice] Manage Progress

25.3.1 Step 1: Define Your Writing Process

Before you can start managing progress, you'll need to identify the high-level phases in your writing process. These provide an overall framework for tracking progress, creating meaningful workflow stages such as Design > Write > Edit > Review > Approve > Publish for each topic and task. For an in-depth discussion about creating a tailored writing process unique to your requirements, see Chapter 7: Tailor the Process.

25.3.2 Step 2: Break Down Scope

Now that you've defined the macro stages of your workflow, you'll need to break down your documentation project's scope into deliverables – or even further, into topics and tasks. We discuss this in more detail in Chapter 10: Make a Plan.

25.3.3 Step 3: Choose Your Progress Tracker

Next, identify the tool that best fits your project's needs for managing progress. Think about your project's complexity, how many writers are involved, and how comfortable you are with various tools. You can either begin from scratch or customize one of our templates to meet the unique needs of your project:

- Technical Writing Process Checklist
- Status Tracker Template
- Sample Documentation Project Schedule

25.3.4 Step 4: Set Up Progress Tracker

After selecting your tool, customize it for your project:

- **Checklists and Status Trackers**: Establish workflow stages and tasks mirroring your writing process.
- **Visual Management Boards**: Adapt your board with columns representing various project phases.
- **Project Schedules**: Align your project schedule with the project's key milestones.

25.3.5 Step 5: Populate With Tasks

Now, it's time to populate your selected tool with tasks. Use the breakdown you developed in your documentation plan as the basis for this. Make sure each task is clear and stands on its own.

25.3.6 Step 6: Manage Progress

Now that you've set up your tracker, it's time to manage it. Regular updates are key. Set a reminder to routinely update the status of your tasks—for example, every week or two, or aligning with sprints if you're working in an Agile development project. If you're leading a team, establish scheduled check-ins to ensure everyone's tasks are progressing well. This practice not only keeps the project moving forward but also encourages team accountability.

> **Insight**
>
> **Ensuring Project Continuity**
>
> Regularly updating your tracker is important for visibility for the rest of the team and your stakeholders, and to maintain continuity. If you need to step away from the day-to-day management of the project for any reason—say, you're on vacation, sick, or called away to work on another higher-priority project—someone else may need to step in to manage your project. Having a regularly updated tracker will facilitate a smooth handover.

25.3.7 Step 7: Manage Issues

"We spent six months documenting a product, and then we were told, 'Thanks for all your hard work; we're going to sunset this product now.' That was literally my first job as a manager. I built a team, and we documented the product, and no sooner had we done it than it was announced it was canceled. Even now, I'm still in disbelief. But it was good practice in learning how to manage a backlog!"

—Robert, Technical Documentation Manager

Even in the best-planned projects, things can go wrong—often through no fault of the technical writer. Regardless, this can be a very frustrating experience. Below, we've listed some common issues that arise in documentation projects, delaying or even derailing them. If you're experiencing one of these issues, make sure to raise it with your manager as soon as possible, as per the next step.

- **Unanticipated Feedback**: Feedback from a subject matter expert who wasn't part of the original review team now needs to be incorporated into your documentation. Similarly, sometimes a member of the review team will take a second look at your documentation (perhaps when they finally have time to review it properly), resulting in a number of fresh, yet poorly timed insights.

- **Product Testing**: Testing sheds new light on a particular feature of the product (for example, it doesn't work as expected), requiring significant changes to the feature, and a corresponding update to the product documentation. Although changes following testing are expected, sometimes the scale of the changes can be substantial.

- **Scope Creep**: Changes to the product scope require adding or removing a new feature, resulting in changes to the product documentation. These changes may be last minute or the result of scope creep: the common phenomenon where a product's scope continues to evolve throughout the development lifecycle. Note that engineering effort is not always proportional to documentation effort. For instance, a one-line code change can completely change a product's user interface, necessitating the retaking of every screenshot in the documentation and generating many hours of rework for the writer.

- **Underestimation/Overestimation**: You've done your planning, carefully breaking down the scope of a new documentation project, only to find that your estimates for something are wildly off—to the point that they impact your overall schedule. Perhaps you didn't fully understand a feature when you sat down to sketch out a draft table of contents, and it turned out to be something much more significant than you thought, or vice versa. Despite your best efforts, predictions can go awry.

- **Organizational Change**: Changes in the organization's systems, processes, or organizational structure occur midway through the project, requiring updates to your process and procedure documentation. Companies are constantly evolving to stay competitive, and these changes can sometimes be abrupt.

- **Project Cancellation**: The product or project gets canceled—and so do your docs. There's nothing more frustrating than laboring on a project, pouring your heart and soul into developing documentation, and going the extra mile to hit critical milestones… only for the project to be canceled, negating the need for your carefully crafted documents. Unfortunately, this does happen!

 Insight

These Issues Sound Risky—What Can I Do About Them?

Do any of the scenarios above actually happen? Absolutely, they do. This section was compiled using feedback from technical writers and documentation managers. If you'd like to see some sample mitigation strategies for the issues described above, check out the Risk Management section in our Documentation Plan Template. Note that in project management jargon, an "issue" is referred to as a "risk" until it occurs.

25.3.8 Step 8: Report Progress

The next step in managing your project's progress is to consistently report your accomplishments and challenges. This communication should be scheduled regularly, such as weekly or biweekly, with key stakeholders including your project sponsor, line manager, or project manager.

Use these meetings as opportunities to demonstrate the value you or your team are contributing. Regularly update on completed work, and involve them in resolving issues by promptly informing them if something is going off track. Don't hide problems! They'll appreciate your transparency and early warnings, setting the stage for a positive working relationship where you proactively collaborate to solve problems.

25.3.9 Step 9: Analyze Data

Status tracking tools are treasure troves of data, filled with insights that shed light on productivity trends for individuals and teams alike. By tapping into the analytics of your status tracking tool, or by integrating additional data fields, you can extract meaningful metrics. These include cycle time (the duration from work start to completion for a workflow stage), work in progress or WIP (the number of tasks being handled at any given time), and review loop interval (the time taken to review documents).

 Insight

Guidance for Analyzing Tracking Data

Our forthcoming book, *Technical Documentation Management*, explains how to use metrics to measure success in your writing projects. You may wish to use these to assist in analyzing and interpreting your tracking data. These chapters will be progressively released to subscribers of Boffin Education's website over the course of 2024.

Templates

The templates in this book have been designed to help you implement your own Technical Writing Process. They're available free to subscribers of this book. Use and adapt them as required to suit your project.

To access the templates:

1. **Subscribe** to the Boffin Education website using the coupon code **TWPROCESS2024**. This will give you one year of free access to all content on the site, including the online version of this book.
2. **Download** the templates in their original Microsoft Office format. They've been created in common Microsoft Office programs: Word, Excel, PowerPoint, and Project.

Here are several helpful tips for using the templates:

- Text in [square brackets] is placeholder text for you to replace with your own version. Don't leave the text in square brackets in your version.
- Text in *italics* is instructions on using the templates. Don't leave the instructions in your version.

The templates have been designed to be as easy to use as possible. However, they do assume a level of basic competence with the Microsoft Office program they were created in.

You may have to adjust the size of tables or spreadsheet columns to suit your requirements.

Templates

List of Templates

This section provides a brief overview of each template.

Template	Short Description
Audience Persona Template	Template for the creation of audience personas with visual avatars in technical documentation projects using the Five Ws method.
Briefing Checklist for Technical Writers	Checklist to guide technical writers in initial discussions in the planning phase, ensuring all essential project aspects are covered.
ChatGPT Prompt Library for Technical Writers	Practical prompts, tips, and instructions for using ChatGPT in technical writing tasks such as research, content planning, writing, editing, and formatting.
Checklist for Using Images in Technical Documentation	Checklist for incorporating images into technical documentation, ensuring effective use of screenshots, diagrams, and photographs.
Documentation Microplan Template	Streamlined planning tool for technical writers, focusing on essential elements for quick and effective documentation planning.
Documentation Microplan Template for Agile Projects	Documentation microplan template for Agile environments, facilitating quick and efficient documentation planning in fast-paced software development projects.

Template	Short Description
Documentation Plan Template	Template for planning technical documentation projects, encompassing all elements from purpose and audience to deliverables, review team definition, timelines, risk management, and more.
Editing Checklist Template	Checklist guiding technical writers through the rigorous editing process of documents. It ensures a consistent and thorough review at each level of editing, from structural edits to proofreading.
Editing Sheet Template	Template for maintaining consistency in style and terminology across technical documents during the editing process.
Resource Checklist for Technical Document Creation	Checklist for sourcing the necessary information prior to commencing a documentation project.
Review Log Template	Template for tracking and organizing feedback received during document review. Categorizes comments from multiple reviewers, helping writers maintain a clear record of changes.
Simplified Technical Writing Process Checklist	Streamlined checklist for managing the progress of documents or topics through the Technical Writing Process. Ideal for keeping track of writing projects.
Status Tracker Template	Excel spreadsheet for monitoring the progress of writing tasks, aiding writers in effectively managing various stages of documentation projects.

Templates

Template	Short Description
Subject Matter Expert (SME) Interview Template	Template for conducting interviews with SMEs. Essential for technical writers in gathering accurate information.
Subject Matter Expert (SME) Workshop Template	Template for facilitating productive workshops with multiple SMEs. It includes scheduling tips, agenda setting, and engagement strategies.
Technical Document Template	Template for creating uniform and professional technical documents, incorporating key elements such as predefined headings and placeholders for document control.
Technical Writing Process Checklist	Streamlined checklist for managing the progress of documents or topics through the Technical Writing Process. Ideal for keeping track of writing projects.
Technical Writing Process Template	Structured guide outlining the core phases and activities for creating technical documentation.
Template Design Checklist	Checklist for creating effective and consistent document templates, ensuring alignment with organizational branding and standards.
Translation Best Practices Checklist for Text and Graphics	Checklist for translating technical documents, including writing and image preparation requirements, for a streamlined and cost-effective process.
Visual Design Principles Checklist for Technical Writers	Checklist for technical writers to apply visual design principles, ensuring their documents are both visually appealing and functionally effective.

Templates

Glossary

This section contains a summary of terms defined in the *Technical Writing Process*. These terms are tailored specifically for the world of technical writing and may have different definitions in other knowledge domains.

Term	Definition
Activity	A task, action, or milestone in a process or schedule. In technical writing, as in process analysis, activities are commonly expressed as verb–noun phrases—"define scope," "write first draft," and so on. Also known as a *task*.
Approval	The formal acknowledgment or sign-off by an approver or document owner that a document is fit for publication.
Approver	An authorizing person within an organization who confirms that a document is fit for publication.
Artificial Intelligence (AI)	The simulation of human intelligence by machines and computer systems.
Audience	Meaning 1: ("An audience") A specific group or groups of people who will interact with or use a document. This can be categorized into primary, secondary, and hidden audiences, depending on their degree of interaction with the document. Meaning 2: ("Audiences") A general term for anyone who interacts with or uses a document, regardless of their categorization.

Glossary

Term	Definition
Benchmark	A point of reference against which the progress, performance, or quality of a technical documentation project can be measured. For example, the average time taken to write a topic of a certain complexity.
ChatGPT	A large language model developed and released by OpenAI, designed to generate humanlike text and content in response to human prompts.
Comments	Written notes about any aspect of a document. These may be specific to a section or paragraph within the document, or general commentary about the document's overall effectiveness. These may be emailed as a summary, affixed to a hard copy with sticky notes, or applied via the comments functionality in an authoring tool or word processor.
Content	The actual material of the document, consisting of text, images, and embedded markup tags. It forms the primary substance and information conveyed in the document.
Contingency	A reserve of time set aside in case something unexpected delays the project, or if the initial assumptions prove too optimistic.
Controlled Document	A document that requires control throughout its lifecycle. It features a unique identifier and approval details. Using an incorrect version of this document could lead to quality, safety, or compliance issues.
Data	Basic units of information, like words, numbers, or images, that represent aspects of reality.

Glossary

Term	Definition
Deliverable	The output of an activity in a process or project. In technical writing, deliverables include user guides, manuals, and procedures—that is, content intended for use by end users—as well as internal documents, such as documentation plans and schedules, that are used in project planning and management.
Dependency	A relationship between tasks where one task depends on another task's start or finish. Examples include finish-to-start, start-to-start, and so on. Also known as a predecessor.
Document	A discrete unit of information used to guide work, decisions, or judgment that serves as a guide to what should be done. Documents are forward-looking, as opposed to records, which are historical. Examples include technical documentation, plans, policies, and engineering drawings.
Document Control	The process of managing documents through the phases of their lifecycle.
Document Owner	A manager or delegate responsible for ensuring a document is current and accurate.
Duration	The total time taken to complete an activity, regardless of the effort (units of labor) involved. See also: *Effort*.
Effort	The units of labor required to complete an activity, typically measured in hours, days, or weeks. For example, "It took me 6 hours (effort) spread over 2 days (duration) to write that document." See also: *Duration*.

Glossary

Term	Definition
External Documentation	User-oriented content for customers and partners outside the organization, emphasizing ease of use, product understanding, and brand representation.
Extrinsic Load	The effort involved in filtering out irrelevant, or "extrinsic," information, as per the field of cognitive psychology.
Feedback	Any comments or markup received from a reviewer in a document review.
Forms	Interactive tools designed for structured data collection, guiding users in providing specific information. Forms can be physical or digital, and are important for organizing data input in an efficient and consistent manner.
Front Matter, Text, and Back Matter	In publishing, content—such as a book—is traditionally divided into several sections: front matter, text, back matter, and a cover. This concept is not just academic—it also applies to many other long-form or hard-copy (i.e., printed and bound) documents, such as reports, plans, and user guides.
Generative Artificial Intelligence	A subset of artificial intelligence models capable of generating new content—such as images, text, or music—that isn't directly copied from training data.
House Style Manual	An organization-specific (or even team-specific) style guide. House style manuals contain guidelines on how to write in the organization's style, dos and don'ts, trademarks, legal boilerplate, the correct tone to adopt, and other relevant details.

Glossary

Term	Definition
Individual Contributor (IC)	Individual contributor (IC) is industry jargon for someone who doesn't manage other people.
Information	Data that has been processed and organized to support decision-making; for example, a document. Remember it as data "in formation."
Internal Documentation	Technical content for an organization's internal audience, focusing on technical details, processes, and operational efficiency.
Intrinsic Load	The effort involved in learning relevant, or "intrinsic," information, as derived from the field of cognitive psychology.
ISO 9001	ISO 9001 is the global standard for quality management systems. It's published by the International Organization for Standardization, a nongovernmental organization that develops and publishes international standards across a wide range of industries and sectors. This standard focuses on ensuring companies meet customer needs for their products and services. It also sets out requirements for document control, which we discuss in depth in Chapter 24: Publish.

Glossary

Term	Definition
ISO 9001	ISO 9001 is the global standard for quality management systems. It's published by the International Organization for Standardization, a nongovernmental organization that develops international standards across a wide range of industries and sectors. This standard focuses on ensuring companies meet customer needs for their products and services. It also sets out requirements for document control, which we discuss in depth in Chapter 24: Publish.
Knowledge	Information that has been internalized by someone, providing the basis for them to act on it.
Large Language Model (LLM)	A subset of generative artificial intelligence models designed to understand and generate humanlike text.
Markup	Symbols such as strikethrough and "red-line" that indicate changes to a document, such as additions, changes, and deletions. These may be written on a hard-copy printout or applied via the track changes functionality in an authoring tool or word processor.
Metadata	Data about data. For example, in digital photography, metadata captured alongside a digital image typically includes the date and time the photo was taken, the brand and model of the camera used, and the geographical location it was taken in.

Glossary

Term	Definition
Metainformation	Metainformation is information about other information. This is a term we've used to describe instructions and placeholder text in templates. In doing so, we've leaned on the definition of a closely related concept, metadata, which is "data that provides information about other data." If you're not clear on the difference between data and information, see Chapter 9: Collect Information: [Theory] DIKW Pyramid.
Milestone	A significant or noteworthy point in your project schedule, such as the approval of a document or the launch of a product. Milestones are typically represented as activities with a "zero" duration.
Personas	A fictional character that embodies the common traits of a specific group of customers or audiences for documentation. These are often referred to as archetypes.
Portfolio	A portfolio is a collection of your own work samples from organizations you've worked for and that you have permission to showcase to prospective employers.
Presentation	The visual styling of the document, including aspects like font size, spacing, color, and layout, making it more user friendly and effective in communication.
Process	A set of activities or tasks performed to accomplish an objective. Processes typically have a trigger that initiates the process; inputs necessary to perform the process; a corresponding result, or output; and a sequence of steps and decision points in between.

Glossary

Term	Definition
Prose Linter	A tool or software application designed to analyze written text for style, grammar, syntax, and sometimes even content consistency. These tools go beyond spell checkers by offering a more in-depth analysis of the text. The concept is borrowed from the world of programming, where a "linter" is a tool that analyzes source code to flag programming errors.
Quality	A measure of how well the work matches generally accepted—but not always stated—standards of good technical writing, including accuracy, clarity, and organization.
Record	A discrete unit of information or collection of data that forms evidence of past activities or decisions, serving as a static memory of what was done. Records are historical, in contrast to documents, which are forward-looking. Examples include invoices, test results, and completed maintenance logs.
Résumé or Curriculum Vitae (CV)	A résumé (also known as a CV, or curriculum vitae) is a summary of your skills, responsibilities and achievements (work experience), and qualifications. Its purpose is to snag you an interview by proving to a prospective employer that you have what it takes to do the job. Think of it as a foot in the door to get you through to the interview stage.
Review	The process of evaluating a document against quality standards such as technical accuracy, consistency with style manuals, templates, branding, and so on.

Glossary

Term	Definition
Reviewer	An expert responsible for evaluating a document against quality standards such as technical accuracy, consistency with style manuals, templates, branding, and so on.
Rhetoric	Rhetoric is the art of effective communication and persuasion. It involves connecting with your audience in a way that encourages them to view things from the speaker's or writer's perspective.
Scope	The sum of the deliverables (like user guides, manuals) or services (such as editing, proofreading) to be provided in a technical documentation project.
Structure	The organization of elements within a document, including headings, sections, and overall layout, which establishes the framework and logical flow of the content.
Style Guide	A manual of guidelines on grammar, punctuation, layout, formatting, structure, and other stylistic aspects of writing. An example is the *Chicago Manual of Style* in the United States. Style guides vary from country to country and across different industries.
Stylesheet	A set of rules governing the visual design of a document, enabling the independent update of presentation elements like fonts, colors, and layout, without affecting the content.
System	An integrated whole consisting of interacting components. These components can be both tangible and intangible, such as the people, processes, knowledge, equipment, and software within a company.

Glossary

Term	Definition
Task	See *Activity*.
Technical Document	A document combining technical information with instructional guidance to help its audience accomplish a goal, such as carrying out a process or using a product. Examples include user guides, developer documentation, procedures, manuals, and quick-reference guides.
Technical Writer	A writer who develops (writes, edits, curates, etc.) technical documents. Also known as a technical communicator or technical author.
Templates	Predesigned frameworks for documents, providing standardized layouts and styles to ensure consistency and efficiency in writing while aligning with an organization's branding and visual identity.
Terminology Database (Termbase)	A collection of standardized terms, often specific to a particular product or industry, used by translators. When these standard terms are encountered in text, their pre-translated versions are automatically inserted by the computer-assisted translation (CAT) tool, ensuring consistent and high-quality translations. Termbases are essential for maintaining term consistency across various translations and are regularly updated and validated for accuracy.
Test Case	A scenario describing a set of steps with a defined start and end point, and an expected result. Test teams can use a test case, along with test data that is sufficiently representative of reality, to verify that a subject, such as a software application and its technical documentation, meets a specific requirement.

Glossary

Term	Definition
Translation Memory	A database used by translation service providers to store completed translations. It is utilized in subsequent translation tasks to identify and automatically translate repeated text, ensuring consistency and reducing the need to retranslate identical content, thereby saving time and costs. Translation memories can be shared between different service providers, facilitating efficiency when multiple vendors are used.
Version	Meaning 1: The specific identifier assigned to a document each time it is updated. This versioning allows for tracking the evolution of a document, as different versions (or revisions) represent the document at various stages of its editing history. Meaning 2: A document's adaptation for different user groups while maintaining the same core content. For example, the same procedure document could exist in two versions: one in English and another in Spanish.
Version Control	A systematic process for tracking and managing different iterations of a document as it is revised and updated over time. This process ensures that all changes made to a document are recorded, enabling an audit trail of its evolution.
Wisdom	The ability of someone to use knowledge, information, and data to make well-informed and ethically sound decisions.

Index

We haven't provided an index for this book because it's designed to be used in conjunction with our searchable online subscription site, https://boffin.education/. Purchasers of the e-book and paperback versions of this book can use the discount code in Templates to obtain a free one-year subscription and access the full breadth of our technical writing content, including forthcoming titles not yet available in print.

References

[1] Gales, C., & Splunk Documentation Team. (2020). *The Product is Docs: Writing Technical Documentation in a Product Development Group* (2nd ed.), p. 121.

[2] Schlotfeldt, J and Bittner, C. (2018, September). How Can You Leverage Data to Know You Have Effective Content? *Intercom*, 65(5), pp. 16-19, at p. 19.

[3] MacKenzie, E. (2022, May 6). *Guide to Technical communication: history, products, skills, education.* MastersinCommunications.com. https://www.mastersincommunications.com/features/guide-to-technical-communication

[4] *The Way to the Stars: Build your own Astrolabe | St John's College, University of Cambridge.* (n.d.). https://www.joh.cam.ac.uk/library/library_exhibitions/schoolresources/astrolabe/chaucer

[5] *Leonardo da Vinci's notebooks · V&A.* (n.d.). Victoria and Albert Museum. https://www.vam.ac.uk/articles/leonardo-da-vincis-notebooks

[6] MacKenzie, E. (2022, May 6). *Guide to Technical communication: history, products, skills, education.* MastersinCommunications.com. https://www.mastersincommunications.com/features/guide-to-technical-communication

[7] Longo, B. (2002). Who Makes Engineering Knowledge? Changing Identities of Technical Writers in the 20th Century United States. *International Conference on Professional Communication (IPCC), Proceedings of IPCC 97.* Communication. https://doi.org/10.1109/ipcc.1997.637031, p. 61.

[8] Longo, B. (2002). Who Makes Engineering Knowledge? Changing Identities of Technical Writers in the 20th Century United States. *International Conference on Professional Communication (IPCC), Proceedings of IPCC 97.* Communication. https://doi.org/10.1109/ipcc.1997.637031, p. 62.

[9] Kassa, D. (2015). *Document Control: Lifecycle and the Governance Challenge.* Unknown Publisher. p. 9

References

[10] Arkin, S. (2022, April 2). *The dos and don'ts of writing samples*. Tested Writing. https://arkinwriting.com/2022/04/02/the-dos-and-donts-of-writing-samples/

[11] *Matplotlib project*. (n.d.). Google for Developers. https://developers.google.com/season-of-docs/docs/2020/participants/project-matplotlib-jeromev

[12] *Salary surveys*. (n.d.). Write the Docs. https://www.writethedocs.org/surveys/

[13] Hackos, J. T. (2007). *Information Development: Managing Your Documentation Projects, Portfolio, and People.* John Wiley & Sons., at pp. 318-325.

[14] Wodecki, B. (2023, March 1). UBS: *ChatGPT may be the fastest growing app of all time*. https://aibusiness.com/nlp/ubs-chatgpt-is-the-fastest-growing-app-of-all-time

[15] Divatia, A. (2023, June 22). *How companies can use generative AI and maintain data privacy*. Forbes. https://www.forbes.com/sites/forbestechcouncil/2023/06/22/how-companies-can-use-generative-ai-and-maintain-data-privacy/

[16] Covert, A. (2014). *How to Make Sense of Any Mess*. Abby Covert, p. 21

[17] Liew, A. (2007, June). Understanding Data, Information, Knowledge And Their Inter-Relationships. *Journal of Knowledge Management Practice*, 8(2).

[18] Rowley, J. (2007). The wisdom hierarchy: representations of the DIKW hierarchy. *Journal of Information Science*, 33(2), 163-180).

[19] *Manifesto for Agile software development*. (n.d.). https://agilemanifesto.org/

[20] Zwikael, O. (2009). *The relative importance of the PMBOK® Guide's nine Knowledge Areas during project planning*. Project Management Journal, 40(4), 94-103, p. 98.

[21] Weiss, J. S. (2022). *PAD Beyond the Classroom: Integrating PAD in the Scrum Workplace* (Doctoral dissertation, University of South Florida).

[22] Weiss, J. S. (2022). *PAD Beyond the Classroom: Integrating PAD in the Scrum Workplace* (Doctoral dissertation, University of South Florida).

[23] Barnhurst, K. G. (2016). *Mister Pulitzer and the spider: Modern news from realism to the digital*. University of Illinois Pres, location 4426.

References

[24] Barnhurst, K. G. (2016). *Mister Pulitzer and the spider: Modern news from realism to the digital*. University of Illinois Pres, location 204.

[25] Kalbach, J. (2020) *Mapping experiences*. O'Reilly Media, pp. 126-127.

[26] Kalbach, J. (2020) *Mapping experiences*. O'Reilly Media, pp. 144.

[27] Rosenfeld, L., Morville, P., & Arango, J. (2015). *Information Architecture: For the Web and Beyond* (Kindle ed.). O'Reilly Media, pp. 23-24.

[28] *The Chicago Manual of Style, 17th ed.* Chicago: University of Chicago Press, 2017. https://doi.org/10.7208/cmos17, s. 1.3 Divisions and parts of a book—overview.

[29] Lambe, P. (2007). *Organising Knowledge: Taxonomies, Knowledge and Organisational Effectiveness*. Chandos Publishing, pp. xv-xvi.

[30] Clark, D. (2007). Content Management and the Separation of Presentation and Content. *Technical Communication Quarterly*, 17:1, 35-60. DOI:10.1080/10572250701588624

[31] Cohen, M. (2004, May 14). Separation: the web Designer's dilemma. *A List Apart*. https://alistapart.com/article/separationdilemma/

[32] Clark, D. (2007). Content Management and the Separation of Presentation and Content. *Technical Communication Quarterly*, 17:1, 35-60. DOI:10.1080/10572250701588624, p. 36.

[33] Williams, R. (2014, 4th ed.). *The Non-Designer's Design Book*. Peachpit Press.

[34] Whitbread, D. (2023, 3rd ed.). *The Design Manual*. Whitbread, Canberra, p. 396.

[35] Kassa, D. (2015). *Document Control: Lifecycle and the Governance Challenge*. Unknown Publisher. Kindle Edition, p. 81.

[36] metadata. (2024). In *Merriam-Webster Dictionary*. https://www.merriam-webster.com/dictionary/metadata.

[37] Miller, G. A. (1956). The Magical Number Seven, Plus or Minus Two: Some Limits on Our Capacity for Processing Information. *Psychological Review*, 63(2), 81.

[38] Cowan, N. (2001). The Magical Number 4 in Short-Term Memory: A Reconsideration of Mental Storage Capacity. *Behavioral and Brain Sciences*, 24(1), 87-114.

References

[39] Quoteresearch. (2011, September 14). *Writing is easy; you just open a vein and bleed – quote Investigator®*. https://quoteinvestigator.com/2011/09/14/writing-bleed/.

[40] Kassa, D. (2015). *Document Control: Lifecycle and the Governance Challenge*. Unknown Publisher. Kindle Edition, p. 79.

[41] Clark, R. C., & Mayer, R. E. (2016). *E-learning and the science of instruction: Proven guidelines for consumers and designers of multimedia learning.* John Wiley & Sons.

[42] Clark, R. C., & Mayer, R. E. (2016). *E-learning and the science of instruction: Proven guidelines for consumers and designers of multimedia learning.* John Wiley & Sons, Fig 4.8.

[43] Mayer, R., & Fiorella, L. (Eds.). (2021). *The Cambridge Handbook of Multimedia Learning (3rd ed., Cambridge Handbooks in Psychology)*. Cambridge: Cambridge University Press. doi:10.1017/9781108894333, Chapter 5.

[44] Sweller, J., Ayres, P., and Kalyuga, S. (2011) *Cognitive Load Theory: Explorations in the Learning Sciences, Instructional Systems and Performance Technologies* (1st ed). Springer. DOI 10.1007/978-1-4419-8126-4, p. 43.

[45] *The Chicago Manual of Style*, 17th ed. Chicago: University of Chicago Press, 2017. https://doi.org/10.7208/cmos17.

[46] Markel, M., & Selber, S. A. (2021). Chapter 13 in *Technical Communication* (13th ed.), pp. 352-363.

[47] Carliner, S., et al. (2014). "What measures of productivity and effectiveness do technical communication managers track and report?" in *Technical Communication*, 61(3), 147-172.

[48] Also known as Directive 2006/42/EC.

[49] *The Globalization Industry Primer*. (2007). Romainmôtier, Switzerland: Localization Industry Standards Association—LISA.

[50] Business News Daily: Small business solutions & inspiration – BusinessNewsDaily.com. (n.d.). *Business News Daily*. https://www.businessnewsdaily.com/.

[51] Kassa, D. (2015). *Document Control: Lifecycle and the Governance Challenge*. Unknown Publisher. Kindle Edition, p. 29.

References

[52] Kassa, D. (2015). *Document Control: Lifecycle and the Governance Challenge*. Unknown Publisher. Kindle Edition.

[53] Cochran, C. (2015). *ISO 9001:2015 in Plain English*. Paton Professional.

[54] Kassa, D. (2015). *Document Control: Lifecycle and the Governance Challenge*. Unknown Publisher. Kindle Edition.

[55] Cochran, C. (2015). *ISO 9001:2015 in Plain English*. Paton Professional, p. 111.

[56] Liker, J. (2020). Chapter "Connect People and Processes Through Continuous Process Flow to Bring Problems to the Surface" in *The Toyota Way, Second Edition: 14 Management Principles from the World's Greatest Manufacturer*. (2nd ed.) McGraw Hill.

[57] Liker, J. (2020). Chapter "A Storied History: How Toyota Became the World's Best Manufacturer" in *The Toyota Way, Second Edition: 14 Management Principles from the World's Greatest Manufacturer*. (2nd ed.) McGraw Hill.

Made in the USA
Middletown, DE
07 July 2025